C0-AOH-824

BIOENERGY – REALIZING THE POTENTIAL

Elsevier Internet Homepage - http://www.elsevier.com
Consult the Elsevier homepage for full catalogue information on all books, major reference works, journals, electronic products and services.

Elsevier Titles of Related Interest

C.J CLEVELAND
Encyclopedia of Energy (6 volume set)
2004, ISBN: 0-12-176480-X

J.R. FANCHI
Energy Technology and Directions for the Future
2004, ISBN: 0122482913

D. TILLMAN & N. HARDING
Fuels of Opportunity: Characteristics and Uses in Combustion Systems
2004, ISBN: 0080441629

J. TRINNAMAN & A. CLARKE
2004 Survey of Energy Resources
2004, ISBN: 0080444105

Related Journals:
Elsevier publishes a wide-ranging portfolio of high quality research journals, encompassing the renewable energy field. A sample journal issue is available online by visiting the Elsevier web site (details at the top of this page). Leading titles include:

Biomass & Bioenergy
Fuel
Bioresource Technology
Fuel Processing Technology
Trends in Biochemical Sciences
Solar Energy
Solar Energy Materials & Solar Cells
Energy Conversion and Management Materials Characterization
Renewable & Sustainable Energy Reviews

All journals are available online via ScienceDirect: www.sciencedirect.com

To contact the Publisher
Elsevier welcomes enquiries concerning publishing proposals: books, journal special issues, conference proceedings, etc. All formats and media can be considered. Should you have a publishing proposal you wish to discuss, please contact, without obligation, the publisher responsible for Elsevier's Energy programme:

Heather Johnstone
Senior Publishing Editor
Elsevier Ltd
The Boulevard, Langford Lane
Kidlington, Oxford
OX5 1GB, UK

Phone: +44 1865 84 3887
Fax: +44 1865 84 3931
E.mail: h.johnstone@elsevier.com

General enquiries, including placing orders, should be directed to Elsevier's Regional Sales Offices – please access the Elsevier homepage for full contact details (homepage details at the top of this page).

BIOENERGY – REALIZING THE POTENTIAL

Edited by

Dr Semida Silveira

2005

ELSEVIER

Amsterdam – Boston – Heidelberg – London – New York – Oxford
Paris – San Diego – San Francisco – Singapore – Sydney – Tokyo

ELSEVIER B.V.
Radarweg 29
P.O. Box 211, 1000 AE
Amsterdam
The Netherlands

ELSEVIER Inc.
525 B Street, Suite 1900
San Diego
CA 92101-4495
USA

ELSEVIER Ltd
The Boulevard
Langford Lane
Kidlington
Oxford OX5 1GB, UK

ELSEVIER Ltd
84 Theobald's Road
London WC1X 8RR
UK

First edition 2005

ISBN: 0-080-44661-2

⊜The paper used in this publication meets the requirements of ANSI/NISO Z39.48-1992 (Permanence of Paper).

Transferred to digital printing 2006

Foreword

Modern bioenergy has gained increased attention in the past decade. Not only does bioenergy provide an effective option for the provision of energy services from the technical point of view, but it is also based on resources that can be utilized on a sustainable basis all around the globe. In addition, the benefits accrued go beyond energy provision, creating unique opportunities for regional development.

Obviously, the potential of deriving energy services from biomass is no novelty, and many countries, including Sweden, have come a long way in developing bioenergy systems. Still, it is not until more recently that the understanding about the far reach of bioenergy options has come to a turning point, and efforts to promote bioenergy started to be made in a more concerted form at a global level. Today, biomass is seen as one of the most promising renewable sources of modern energy services in the medium term.

In fact, studies about the global biomass potential have multiplied in the past years, contributing significantly to the recognition of the merits of bioenergy beyond expert fora. Markets for bioenergy-related products have grown fast, denoting changes on the demand side, and increasing business interest in the area. This motivates new questions, for example about the need to standardize bioenergy products. It requires renewed attention from other industries that also depend on biomass resources, and demands new types of policies to promote bioenergy which are sensitive to the interests of various industries.

Thus the challenges around bioenergy are many. The development of bioenergy systems with the reliability required of modern energy systems involves sustainable natural resource management, sophisticated organization schemes, and proper market strategies under competitive energy markets. Despite the progress attained in many countries, these challenges should not be underestimated particularly when a broad use of bioenergy is contemplated, not least in less developed countries where energy needs are still very large.

This discussion on potential and challenges has motivated the ***International Workshop on Biomass Potential and Utilization in Europe and Developing Countries***, held in Eskilstuna a couple of years ago, and organized by the Swedish Energy Agency in collaboration with the Swedish International Development Assistance Agency. On that occasion, the Swedish experience served as a starting point for discussing bioenergy solutions for heat and power in particular. However, the success of that meeting lay not only in the interest shown to Swedish solutions but, most of all, also in the variety of contributions and possible solutions that were presented, which emphasized once more the large spectrum of the bioenergy options available for further exploration.

The workshop in Eskilstuna should be seen as part of a range of activities that envisage the promotion of bioenergy utilization. The objective was not to make a comprehensive review of initiatives or to rank them in any particular fashion but to discuss models, opportunities and difficulties that need to be addressed. This publication compiles some of the contributions brought to Eskilstuna, reproducing questions and solutions discussed. We hope that the book will serve as a source of information and inspiration to policy makers, financiers, developers and companies that are in the position to explore bioenergy as a new business in their sphere of activities.

The information provided here offers a starting point for understanding the complexities involved in deploying biomass energy options but, most of all, it serves as a channel to communicate that effective solutions are possible and are being implemented at various scales and under different social, economic and technical conditions. It should be seen as a discussion forum for evaluating existing options and discussing relevant policies and measures that will shape bioenergy utilization in Europe and other regions of the world, as well as to provide ideas for the direction that research should take to support the deployment of bioenergy.

Lars Tegnér
Director of Development
Swedish Energy Agency

About the Authors

Brew-Hammond Abeeku is the Director of the Kumasi Institute of Technology and Environment (KITE) in Ghana and the Managing Director of the Kumasi Energy (KE) Company Limited. He is a mechanical engineer and holds a PhD in Science and Technology Policy from the University of Sussex. He is also a senior lecturer at the Kwame Nkrumah University of Science and Technology in Kumasi, Ghana.

Lars Andersson is a senior project manager at the Swedish Forest Administration and head of international cooperation at the Regional Forestry Board of Värmland-Örebro. He has been forestry advisor and consultant in the Programme for an Environmental Adapted Energy System in the Baltic countries, and in various other programs in the Baltic Sea region such as Baltic 21 and the bioenergy group under BASREC.

Ausilio Bauen is a PhD research fellow at the Imperial College's Centre for Energy Policy and Technology and head of the BioEnergy group at ICEPT. He has researched and consulted extensively on technical, economic, environmental and policy issues relating to decentralized generation and alternative fuel production and infrastructure. His recent focus is on biomass, fuel cells and integration of renewables into energy systems.

Knut Bernotat is a civil engineer in industrial economics and management from the Technical University, Darmstadt, Germany and the Royal Institute of Technology, Stockholm, Sweden. He also holds an international masters degree in environmental engineering and sustainable infrastructure.

Oscar Braunbeck is a full time research, teaching and extension associate professor at the State University of Campinas, São Paulo, Brazil. His focus is on machine design (simulation and optimization), mostly related to forage and sugarcane harvesting. He has coordinated or participated in the design of approximately fifteen field equipments aimed at increasing sustainability in the use of biomass resources.

Luís A.B. Cortez is an Associate Professor at the School of Agricultural Engineering at the State University of Campinas – UNICAMP in Brazil. He is an agricultural engineer and received his PhD in Engineering from Texas Tech University, USA in 1988. Since then he is working in the field of energy in agriculture with emphasis in biomass conversion.

Upali Daranagama is a chemical engineer with 25 years of experience in the energy sector. Presently, he serves as the United States Agency Development – USAID – Colombo office as project management specialist in energy

Dominic N. Derzu is a senior project engineer at Kumasi Energy (KE) Company Limited, Ghana. He is a mechanical engineer, who also had an industrial training in France on industrial equipment and technologies. In the last two years, he has been involved in biomass project development at the Kumasi Institute of Technology and Environment (KITE).

André Faaij is an Assistant Professor of Energy Supply and Systems Studies at the Copernicus Institute, Utrecht University. He has a background in chemistry and environmental sciences with a PhD in bioenergy. He has done advisory work to FAO, IEA and UN among others and published extensively. He is task leader of IEA task 40 on Sustainable Bioenergy trade, lead author for the World Energy Assessment and the IPCC, and coeditor of Biomass & Bioenergy.

Marco Antonio Fujihara is an agricultural engineer specialized in forest economy. Presently, he is a senior consultant on climate change at PriceWater House Coopers. In the past years, he has worked with the certification of energy and forest companies and developed CDM projects, actively making climate change an important issue in the managerial agenda of companies and governments. He was previously director at the Brazilian Institute of Environment and Renewable Natural Resources.

Luiz Carlos Goulart is an environmental manager at Plantar S.A., Brazil. He is a business administrator and specialist in finance, environment and quality systems. In recent years he has worked full time with the Plantar CO_2 Project, being responsible for all basic documents of the project such as Baseline Study, Monitoring and Verification Protocol, Financial Due Diligence, Environmental Assessment and other small reports and researches.

Marlies Härdtlein is a project manager at the Institute of Energy Economics and the Rational Use of Energy, University of Stuttgart. She elaborated her doctoral thesis in the field of sustainable production and utilization of energy crops. Currently, she coordinates a project on standardization of solid biofuels and conducts research on economic and ecological aspects of biofuel provision and utilization, as well as quality assurance.

Marcelo Junqueira is an agricultural engineer and holds a masters degree in business administration. He has many years of experience in the sugar industry in Brazil. In 2002, he co-founded Econergy Brasil, the representative of Econergy International Corporation, where he is now developing CDM and renewable energy related projects in Brazil. He developed the first baseline methodology analyzed by the CDM Executive Board.

Martin Kaltschmitt is the managing director of the Institute for Energy and Environment in Leipzig, Germany. He has been deeply involved in biomass research for more than a decade. Within this time, together with others he published several books as well as some 100 articles and conference contributions in this field. He is also a member of the CEN Technical Committee 335 "Solid Biofuels".

Alexandre Kossoy is the project manager of the Carbon Finance Unit at the World Bank where he carries out financial due diligence of projects and companies in Asia and Latin America. Previously, he worked for the Rabobank International in São Paulo, where he was responsible for the first structured commercial loan for a Kyoto Protocol related project (USD 5 million to the Plantar Project).

Erik Ling is the Business Development Manager for Biofuels at Sveaskog, Europe's largest corporative forest owner. He has a PhD in forest economics. His research focused on institutional issues and different aspects of competitiveness of bioenergy. Previously, he was an executive officer at the Swedish Energy Agency, dealing with forest carbon, bioenergy systems analysis and standardization issues.

Isaías Macedo works at the State University of Campinas (UNICAMP), São Paulo, Brazil. He is a mechanical engineer and holds a PhD in thermal sciences. Until 1982, he worked as a professor at ITA and UNICAMP and did research mainly in energy generation systems. For twenty years, he was at the Copersucar Technology Center, São Paulo, leading research in sugarcane production and processing to sugar and energy.

Geraldo Alves de Moura is the Director and a shareholder of Plantar S. A. He is the leader of the Plantar CO_2 team and is responsible for the companies' climate policies. He has successfully conducted the negotiation of the Plantar Project with the Prototype Carbon Fund at the World Bank which was a landmark in carbon credit financing in the forestry and metallurgical sector.

Christian Rakos has studied physics, philosophy and history in Vienna. Between 1986 and 1997 he worked at the Austrian Academy of Science in the Institute for Technology Assessment. Since 1997, he is working for E.V.A, the Austrian Energy Agency, where he is responsible for renewable energy issues. His focus is on the use of renewable energy in the heat market.

Kamal Rijal is a Sustainable Energy Policy Advisor of the Bureau of Development Policy, UNDP, Bangkok. Prior to this, he served as a renewable energy specialist at the International Centre for Integrated Mountain Development (ICIMOD). The focus of his work at UNDP is on mainstreaming energy issues for poverty alleviation and environmental sustainability towards sustainable human development.

Thomas Sandberg is a Professor of industrial economics and management at the Royal Institute of Technology, Stockholm, Sweden. He is a social scientist and holds a PhD in business studies. After working with organizational issues for many years, he is now specialized in local energy systems.

Semida Silveira is sustainability expert at the Swedish Energy Agency. She has a PhD in regional planning from the Royal Institute of Technology and has done research at institutions such as MIT and IIASA. She is a senior lecturer at the Royal Institute of Technology, and has previously worked as a manager for climate and energy programs at the Stockholm Environment Institute, and as a consultant in environmental business development.

Mônica R. Souza is a mechanical engineer and holds a PhD in energy planning. The focus of her research is on electricity production from biomass. She has worked for almost two years as a researcher at the Utrecht University, The Netherlands.

Daniela Thrän is a project manager at the German Institute for Energy and the Environment. She is an environmental engineer and holds a PhD in civil engineering from the University of Weimar. She coordinates projects in the biomass and renewable energy field and is involved in the European and international bioenergy activities on standardization and quality assurance.

Arnaldo Walter is an Assistant Professor at the Department of Energy at the State University of Campinas, UNICAMP, Brazil. His professional interests include energy planning and technical–economic analysis of energy systems. In recent years, his focus has been on the analysis of electricity production from biomass.

Priyantha Wijayatunga is the Director General of the Public Utilities Commission which is responsible for regulating the restructured electricity industry in Sri Lanka. He is also a professor in electrical engineering at the University of Moratuwa and was previously Dean of the faculty of IT. He has a PhD in power system economics from the Imperial College.

Contents

Foreword – ***Lars Tegnér*** v

About the Authors vii

PART I EXPLORING THE BIOENERGY POTENTIAL 1

1 How to Realize the Bioenergy Prospects? 3
Semida Silveira
1.1. What is the News? 3
1.2. This Book 5
1.3. Bioenergy as Part of the Renewable Basket 7
1.4. The Turning Point 12
1.5. Taking the Leap towards Bioenergy 15
References 16

2 Biomass in Europe 19
Ausilio Bauen
2.1. Is Biomass Important to Europe? 19
2.2. Biomass Resources and Conversion Technologies 20
2.3. The Role of Biomass in Climate Change Mitigation 21
2.4. The EU Energy and Agriculture Policies 24
2.5. Examples of Country Policies within the EU 26
2.6. Concluding Remarks 29
References 30

3 New Challenges for Bioenergy in Sweden 31
Erik Ling and Semida Silveira
3.1. Bioenergy in Transition 31
3.2. Biomass Utilization in Sweden 31
3.3. Important Drives Affecting Bioenergy Utilization 35
3.4. Four Major Tasks in the Development of Bioenergy in Sweden 39
3.5. Concluding Remarks 45
References 46

4 Dissemination of Biomass District Heating Systems in Austria: Lessons Learned ... 47
Christian Rakos
4.1. District Heating in Austria ... 47
4.2. The Diffusion of BMDH in Austrian Villages ... 48
4.3. Technology Performance and Qualification of Professionals ... 48
4.4. The Socioeconomic Conditions of Villages ... 51
4.5. Economic Aspects of Plants ... 52
4.6. The Sociocultural Context ... 54
4.7. The Role of Policies in Supporting Technology Introduction ... 56
4.8. Conclusions ... 56
References ... 57

PART II MANAGING RESOURCES AND ENHANCING BIOMASS PRODUCTION ... 59

5 Managing Fuelwood Supply in Himalayan Mountain Forests ... 61
Kamal Rijal
5.1. The Importance of the Forest Sector in Mountain Areas ... 61
5.2. Energy Services in the Hindu Kush Himalayan Region ... 62
5.3. Fuel from Mountain Forests ... 65
5.4. Major Issues Pertaining to Fuelwood ... 68
5.5. Future Directions for Wood Energy Development in the HKH Region ... 69
References ... 72

6 Modernizing Cane Production to Enhance the Biomass Base in Brazil ... 75
Oscar Braunbeck, Isaías Macedo and Luís A.B. Cortez
6.1. Biomass Availability can be Enhanced in Brazil ... 75
6.2. The Sugarcane Industry as Energy Producer ... 76
6.3. Research and Technology Development in Sugarcane Agriculture ... 78
6.4. From Cane Burning to Mechanical Harvesting ... 80
6.5. Towards Mechanized Green Cane Harvesting in Brazil ... 81
6.6. Trash and Bagasse – Same Source but Different Features ... 86
6.7. Using Trash and Bagasse for Energy Purposes in Different Industries ... 90
6.8. Realizing the Biomass Potential in the Sugar–Ethanol Segment ... 92
References ... 92

7 Integrating Forestry and Energy Activities in Lithuania Using Swedish Know-how 95

Semida Silveira and Lars Andersson

7.1. Bilateral Cooperation for Know-how and Technology Transfer 95

7.2. Forest Management in Lithuania 96

7.3. Fuelwood Utilization in Lithuania 98

7.4. Demonstration Projects in Rokiskis Forests 100

7.5. New Technologies and Management Practices for Higher Productivity and Reduced Costs 100

7.6. Continuing Efforts in the Baltic Sea Region 105

References 108

PART III PROMOTING BIOENERGY UTILIZATION 111

8 Potential for Small-scale Bio-fueled District Heating and CHPs in Sweden 113

Thomas Sandberg and Knut Bernotat

8.1. Aiming at Sustainable Energy Systems 113

8.2. A Method to Estimate the Heat Demand 114

8.3. Potential for Small-scale District Heating and CHP in a Small Region 116

8.4. Potential for Small-scale District Heating in the Counties of Kalmar, Örebro and Västernorrland 119

8.5. The Potential for Small-scale District Heating and CHP in Sweden 120

8.6. The Benefits 123

References 124

9 Cofiring Biomass and Natural Gas – Boosting Power Production from Sugarcane Residues 125

Arnaldo Walter, Mônica R. Souza and André Faaij

9.1. Why Cofiring? 125

9.2. The Rationale 126

9.3. Cases and Hypotheses for Simulation 128

9.4. Simulation and Feasibility Results 131

9.5. Comparison of Alternatives 137

9.6. Final Remarks 138

References 139

10 Techno-Economic Feasibility of Biomass-based Electricity Generation in Sri Lanka 141
Priyantha Wijayatunga, Upali Daranagama and K.P. Ariyadasa
10.1. Introduction 141
10.2. Land Availability 141
10.3. Energy Plantations in Sri Lanka 144
10.4. Technology Options 145
10.5. Economic Analysis 146
10.6. Conclusions 149
References 150

11 Classification of Solid Biofuels as a Tool for Market Development 153
Daniela Thrän, Marlies Härdtlein and Martin Kaltschmitt
11.1. The Need for a Solid Biofuel Standardization 153
11.2. What should be Standardized? 155
11.3. Building a Solid Biofuel Standardization Practice in Europe 157
11.4. Quality Assurance – Example of Straw Quality Improvement 161
11.5. Final Remarks 164
References 164

PART IV EXPLORING OPPORTUNITIES THROUGH THE CLEAN DEVELOPMENT MECHANISM 167

12 The Clean Development Mechanism (CDM) 169
Semida Silveira
12.1. The Challenge of Mitigating Climate Change 169
12.2. The Concept of CDM 170
12.3. The CDM Project Cycle and Institutional Framework 172
12.4. Who will Participate in CDM Projects and Why? 174
12.5. CDM and Bioenergy Options 176
References 178

13 The Role of Carbon Finance in Project Development 179
Alexandre Kossoy
13.1. Introduction 179
13.2. Risk Analysis versus Pricing 180
13.3. Carbon Rights, the Emission Reductions Purchase Agreement (ERPA) and Risk 182

13.4. Examples of Carbon Finance Leveraging Private and Public Investment in Projects from Different Sectors and Countries 183
13.5. Conclusions 186

14 Cultivated Biomass for the Pig Iron Industry in Brazil 189
Marco Antonio Fujihara, Luiz Carlos Goulart and Geraldo Moura
14.1. The Plantar Project 189
14.2. Overview of the Pig Iron and Steel Sectors in Brazil 190
14.3. Baselines 193
14.4. Project Boundaries and Leakage 194
14.5. Environmental Issues 196
14.6. Socioeconomic Issues 197
References 198

15 Carbon Credits from Cogeneration with Bagasse 201
Marcelo Junqueira
15.1. The Context of Santa Elisa's Bagasse Cogeneration Project 201
15.2. Cogenerating with Bagasse – The Project Milestones 202
15.3. Additionality 204
15.4. Project Baselines 205
15.5. Quantifying Baseline Carbon Intensity 207
15.6. Carbon Accounting Evaluation Methods 209
15.7. Lifetime of the Project 211
References 212

16 Wood Waste Cogeneration in Kumasi, Ghana 213
Dominic Derzu, Henry Mensah-Brown and Abeeku Brew-Hammond
16.1. The Increasing Energy Demand in Ghana 213
16.2. Availability of Wood Wastes in Ghana 214
16.3. Feasibility of a Cogeneration Project in Kumasi 214
16.4. Boundary and Baseline of the CDM Project 216
16.5. Certified Emission Reductions (CERs) 217
16.6. Outstanding Issues 218
References 219

PART V MEETING THE CHALLENGES AND MAKING A DIFFERENCE 221

17 Bioenergy – Realizing the Potential Now! 223
Semida Silveira
17.1. Beyond the Barriers to Bioenergy Utilization 223
17.2. Finding Common Ground to Understand and Deal with Trade-offs 224

17.3. Combining Policies, Technology and Management to Develop Innovative Markets 227
17.4. Global Solutions Need Local Solutions – Implementing Strategies for Sustainable Development at Project Level 229
17.5. Mobilizing Forces Towards Sustainable Energy Systems 234
References 235

Index 237

Part I

Exploring the Bioenergy Potential

Chapter 1
How to Realize the Bioenergy Prospects?

Semida Silveira

1.1. WHAT IS THE NEWS?

Biomass has been a major source of energy in the world since the beginning of civilization. It has been important in development processes, including early stages of industrialization in several countries. In Sweden, for example, the first concerns about preservation date from the seventeenth and eighteenth centuries, resulting from the recognition of the central role played by forests in energy provision (see also Kaijser, 2001). Biomass was also essential in the initial development of the iron industry in Sweden and, later on, the same happened in Brazil, where charcoal is still largely utilized in iron reduction. Biomass remains a major source of energy in many countries. Ethiopia and Tanzania, for example, derive more than 90 per cent of their energy from biomass. In fact, the African continent as a whole relies heavily on biomass resources for the provision of energy services.

When observing what happened in the past two centuries, we have the impression that the more industrialized a country became, the more dependent it grew on fossil fuels. But there are exceptions. Norway, for example, was able to industrialize without developing the typical dependency on fossil fuels thanks to its hydropower endowments. At a global level, however, the industrialization period has been characterized by an increasing use of fossil fuels as energy carriers. Thus there is a tendency to think that countries with large biomass dependency are poor countries with a low level of industrialization. The generalized view has been that countries climb an energy ladder that leaves biomass behind in favor of more efficient fuels and technologies, which are often based on coal, oil and gas.

In the past decades, the old rule, that the richer and the more industrialized a country is, the more dependent it becomes on fossil fuels, has been broken. Many countries have realized the need to harness local resources to increase the security of energy supply, reverse fossil fuel dependency and improve trade balance. The global environmental agenda, for example in the form of the Agenda 21 and the Climate Convention, has also played a role in this process for more than ten years now. As a result, there is a general trend to search for energy alternatives involving locally available renewable resources, while simultaneously pursuing increased energy efficiency throughout the economy. Countries have chosen different paths to move towards sustainable energy systems, and the accomplishments vary significantly.

The good news is that the connection often made between biomass utilization and poverty starts fading. All types of energy services can and are being provided today using biomass, with the reliability, safety and efficiency required by the modern economy and society. Moreover, this is not only happening in rich countries, it is also happening in many developing countries. The other good news, and part a corollary of the former, is that industrialization, which is seen as an important step in the development process, can be achieved using sources of energy other than fossil fuels, and this can create jobs and contribute to regional development instead of displacing people, eroding local economies and destroying the natural environment.

There are reasons to believe that the turn of the century has also been a turning point for bioenergy. This results not only from the recognition of the bioenergy potential, but also from the maturity of technologies, the reliability of positive results achieved so far, and the awareness of policy makers about the multiple benefits accrued from bioenergy. To developing countries, this means that the old idea of climbing an energy ladder that gradually goes from biofuels to fossil fuels as a way to access modern energy services should be questioned and reviewed under the light of recent technological development and international opportunities for investing in renewable alternatives.

This may sound almost like a manifesto for bioenergy. Let it be so. Biomass can be used to produce different forms of energy such as heat, electricity and transport fuels, thus providing all the energy services required in modern society. We know that. Some countries have actually come a long way in testing technologies and models that can be replicated. These countries are already realizing their biomass potential. In Sweden, for example, biomass already accounts for 16 per cent of the total energy supply. In Finland, biomass responds to 19 per cent of the country's total supply. In Brazil, 27 per cent of the energy comes from biomass, almost half the part being sugar-cane based, including an annual production of some 10 million m^3 of ethanol which are used in the transport sector. In these countries, biofuels are being used to feed modern and efficient systems, providing essential energy services.

Truly, opportunities come with challenges. We have to face the crude fact that, despite all efforts being made to introduce renewables and despite their rapid percentual growth in many regions, fossil fuel annual additions to the world energy supply are still much larger in absolute terms. A quick look at OECD countries reveals that most of them still depend about eighty per cent or more on fossil fuels for the provision of energy services. Also, developing countries are largely meeting their increasing energy demands with fossil fuels, thus replicating past trends and nonsustainable experiences. Unless very significant and more proactive measures are taken both nationally and internationally, this situation will persist for many years to come, delaying the shift towards sustainable energy systems.

Bioenergy options are at hand, satisfying technical, commercial, environmental, social and even political requirements. Energy infrastructure is important for social and economic development in modern societies, and bioenergy is attractive at all stages of development due to its potential integration with development strategies in rich and poor countries alike, and in comprehensive ways hardly matched by other alternatives. It is no exaggeration to see bioenergy options amongst the most attractive energy forms that we can harness today, with technologies and system solutions that are already mastered, with strong public and political acceptance, and often also with a commercial appeal.

Certainly, we ought to be realistic about what can be accomplished, and at what speed and range. A sustainable use of biomass requires comprehensive management of natural resources such as land and water. There are a number of factors that need consideration when it comes to achieving a fair balance in the use of scarce resources. For example, it is necessary to guarantee that land competition does not jeopardize food production and security. In addition, there are questions of security of supply, vulnerability of energy systems and the challenging task of designing policies that can address the development of multisector systems. Still, these broad tasks should not keep us away from ambitious targets, particularly in the face of promising multiple rewards in the direction of sustainability.

1.2. THIS BOOK

In this book, we talk about opportunities and challenges when it comes to harnessing the biomass potential. In other words, we consider ways through which bioenergy can contribute to global sustainable energy systems. What types of energy services can be provided and how? What needs to be addressed when implementing bioenergy systems? We do not try to be comprehensive but we do move about in the different realities of Europe and developing countries, where needs and demands can mean different things though the bioenergy benefits can be quite similar. We address resource management, markets, technological and institutional development, and policy issues.

Our major task is to show accomplishments and indicate possible directions. We also provide views of different stakeholders so that we can better understand their concerns and the specific roles they can play in the implementation of solutions. We are not proposing a business plan, but we are perhaps suggesting that we should be working on a strategic plan from which various business plans can be generated. Why do we need a broad framework to move forward? – Because the tasks are many, and the potential impacts, very significant. There is need for a multisectoral coordination of action, and that requires appropriate plans for timely and speedy moves, which prove effective in both short and long terms.

We are asking questions such as what the main forces enhancing bioenergy utilization are. Where and how are opportunities being sized, and how can ongoing initiatives be enhanced? What policies are being applied to foster biomass technologies, and how can they be improved? How can the environmental and social benefits of bioenergy be better highlighted and valued, in order to increase the bioenergy attractiveness? What are the challenges ahead and how should they be framed towards effective action? Thus, we are now beyond the question of whether biomass is an attractive and effective energy carrier. Our focus is not on the problems, but on the opportunities. We identify demands and questions related to next steps in developing bioenergy systems, and try to answer some of them by indicating possible solutions.

This chapter provides an introduction to the role of bioenergy in perspective and as it stands today, and a discussion of how bioenergy prospects can be realized and framed towards sustainable development. Throughout the book, the demands and prospects are further discussed, the role of accumulated knowledge and experience reviewed, and new tasks identified.

Chapters 2 to 4 explore policies to promote bioenergy utilization. In Chapter 2, Bauen discusses the policy framework in the European Union (EU), and gives examples of how these policies are reflected in national action. Chapter 3 describes the Swedish experience, and ways through which bioenergy utilization can be enhanced in the country in the near future. In Chapter 4, Rakos addresses the issue of public acceptance to the introduction of district heating systems in Austrian villages, providing a concrete case to exemplify the effect of policies and technology dissemination at the local level.

Chapters 5 to 7 are focused on the management of biomass resources and enhancement of biomass production. Rijal addresses relevant issues in the context of Himalayan Mountain Forests while Braunbeck et al. discuss ways to enhance the biomass base of cane production in Brazil. In Chapter 7, Silveira and Andersson discuss the integration of forestry and energy activities in Lithuania where Swedish experiences are providing the know-how basis for helping explore the local bioenergy potential.

Chapters 8 to 11 discuss ways to promote bioenergy utilization. Sandberg and Bernotat assess the potential for new district heating systems in three counties in Sweden. In Chapter 9, Walter et al. discuss how cofiring natural gas and biomass can be an interesting alternative both technically and commercially. Wijayatunga et al. show a feasibility study done for Sri Lanka where biomass is contemplated as an alternative for the provision of electricity. In Chapter 11, Thrän et al. describe the international work within the EU aimed at standardization of biofuels as a tool to boost markets. These chapters are particularly relevant for the methodologies they present.

Chapters 12 to 16 discuss the Clean Development Mechanism (CDM) to the Kyoto Protocol as a means of promoting bioenergy projects. Economic advantages, development priorities and climate change mitigation are addressed in Chapter 12. The emphasis, however, is given to aspects of project implementation. In Chapter 13, Kossoy discusses the CDM in a business context, particularly from a financial point of view. Chapters 14 to 16 provide examples of CDM projects in Brazil and Ghana.

Chapter 17 closes the book with some final considerations on the trade-offs involved in the choice of energy options, and the need for comprehensive strategies and systems integration to achieve the sustainability goals. Some considerations are also made about the platforms available for enhancing synergies and the ultimate value of energy projects. How can so many opportunities be combined effectively towards the realization of the bioenergy potential and sustainable development? What role can the developing countries play at a global level?

1.3. BIOENERGY AS PART OF THE RENEWABLE BASKET

IEA (2003) estimates that 13.5 per cent of the total 10 038 Mtoe of primary energy supply in the world came from renewable sources in 2001. As much as 79.6 per cent came from fossil fuels, and 6.9 per cent came from nuclear power. Over the last thirty years, the average increase in the utilization of renewables went hand in hand with the increase in energy supply, or around 2 per cent per year (IEA, 2002). Unfortunately, this implies a faster absolute increase in the use of fossil fuels. In fact, the absolute use of fossil fuels increased five times more than the use of renewables in the last three decades.

Since 1990, the primary energy supply in the world grew by 1.4 per cent per year while the growth of renewables was 1.7 per cent per year, indicating not only a slower increase in the use of energy but also a slightly more rapid increase in the use of renewables when compared with other sources. Nevertheless, fossil fuel utilization is still increasing faster in absolute terms as renewable sources are still at low levels. Thus much remains to be done in order to shift world energy systems towards sustainable solutions.

Figure 1.1 illustrates the shares of various renewables in the world energy supply. New renewables such as solar, wind and tide comprise a very small fraction, corresponding to less than 0.1 per cent of the total energy supply of the world and only 0.5 per cent of the renewables. Biomass is by far the most significant renewable source, representing 10.4 per cent of the world total. It is worth pointing out

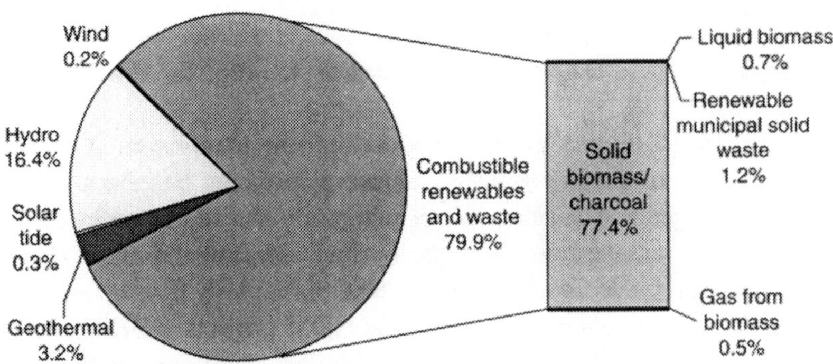

Figure 1.1. World renewable energy supply by source, 2001. Source: IEA (2003).

that while 87 per cent of the biomass resources are used in developing countries, 86 per cent of the new renewables are found in OECD countries (IEA, 2002). In any case, given the small amounts of the latter, developing countries are, in fact, much larger users of renewables than industrialized nations. In addition, it is important to remember that, though making a relatively small contribution to the world's total supply, renewables allow energy to arrive at remote and isolated locations, thus often making a crucial contribution.

Biomass is mostly used in solid form and, to a lesser extent, also in the form of liquid fuels, renewable municipal solid waste and gas. However, recent trends show a faster increase in the use of liquid biomass and municipal waste than solid biomass. In fact, when compared with other renewables, solid biomass showed the slowest growth since 1990. While solar and wind energy supply grew by 19 per cent, solid biomass grew by only 1.5 per cent per year during the 1990s. On the other hand, non-solid biomass and waste such as municipal solid waste, biogas and liquid biomass grew by 7.6 per cent per year. Thus some opportunities are being sized particularly as a result of efforts to find new alternatives to fossil fuels in the transport sector and in waste management. Nevertheless, considering the resource base that is readily available and the great potential to grow biomass, there is much more that can be done to enhance the role of bioenergy.

In the so-called *rich and green scenario* developed by IIASA/WEC, biomass could account for 20 per cent of the total amount of the world energy in 2100 (Nakicenovic et al., 1998). Obviously, this will not happen by itself, and the slow growth of solid biomass provides an illustration of that. This scenario includes significant technological progress and strong international cooperation around environmental protection and equity issues. It is also important to point out that biomass utilization in the IIASA/WEC scenario differs from the present conditions especially when it comes to technology. In particular, significant changes in the way biomass is being utilized in many developing countries today will have to be accomplished. We are basically talking about going from traditional to modern and efficient technologies that can provide high-quality energy services, many of which require access to electricity.

There are significant regional differences when it comes to the availability and use of biomass resources in the world (see Figure 1.2). In many regions of developing countries, biomass is the only accessible and affordable source of energy. In Africa, for example, biomass corresponds to half of the total energy supply. Most of the biomass used in the continent is being harvested informally and only a small part is commercialized, with biomass markets usually operating in urban areas only. In many parts of Asia and Latin America, on the other hand, modern and commercial bioenergy options are readily available and significant. The Brazilian ethanol programme is noteworthy as the single most important accomplishment in providing an alternative fuel to the transport sector.

In addition to woodfuels, other biomass fuels such as forest and crop residues as well as animal waste are common sources of bioenergy in poor countries, where also traditional technologies predominate. Besides the amount of biomass that is readily available in the form of residues, and the potential for improved efficiency in technologies being presently applied, many countries still have land available for energy plantations. Integrating biomass harvesting for energy purposes with forestry and agricultural activities is another option. In many regions, the use of biomass still needs to become sustainable, this being true both where traditional and modern technologies are applied.

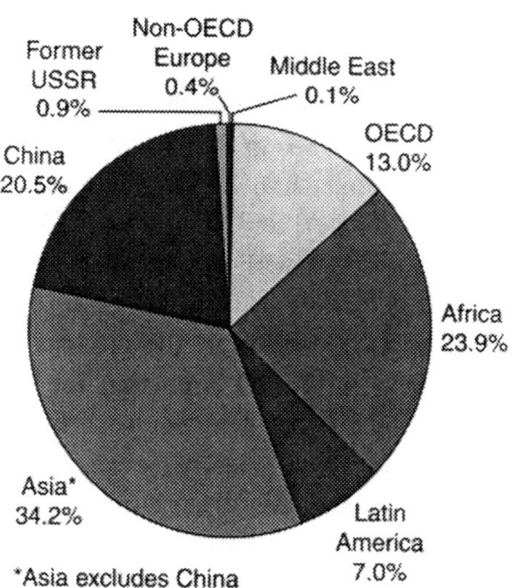

Figure 1.2. Regional shares of bioenergy supply. Source: IEA (2003).

Figure 1.2 shows how the utilization of biomass is distributed across the globe. What it does not say, however, is how large the actual potential for harvesting biomass resources is in the various regions. In fact, the most promising areas are found in the tropical regions. The best average yields per hectare have been observed in sugarcane plantations in Zambia which have reached 1350 GJ/ha/year (global average 650 GJ/ha/year), followed by best-performing eucalyptus plantations in Brazil with 1000 GJ/ha/year (Brazilian average 450 GJ/ha/year). For comparison we can mention that registered US record yields for maize are slightly over 400 GJ/ha/year while the average is about half, and the high estimates from American commercial forests are less than 100 GJ/ha/year (IPCC/SAR, 2001).

A large part of the biomass in developing countries is used in households for cooking and heating. But biomass is also an important energy source in many industries, for example, in the production of ceramics and beverages, and in drying and processing food. These same industries provide an important demand base and starting point for realizing bioenergy projects in developing countries, not least integrated with other established commercial activities. These opportunities are often forgotten for reasons such as lack of knowledge of how to develop bioenergy systems, nonexistence of supporting policies, lack of managerial capacity and conventional energy planning practices.

Only 13 per cent of the total biomass is consumed in the OECD countries, where it accounts for some 3 per cent of the energy supply. In fact, renewables as a whole correspond to only 5.7 per cent of the total primary energy supply in OECD countries, of which about half is being used to generate electricity. The use of solid biomass has had a positive development in OECD countries, showing an annual increase of 1.8 per cent since 1990 as opposed to 1.5 per cent in non-OECD countries. As previously observed, the segments utilizing municipal solid waste and producing liquid biomass are the ones growing faster. While wind and solar energy have reached growth rates higher than 20 per cent per year, liquid biomass has grown at an annual rate of 84 per cent in the OECD. Certainly, all these large growth rates have to be considered with caution as the starting points for renewables have been quite low.

It is also worth noticing that, although the electricity demand is growing by more than 2 per cent per year in OECD countries, the electricity generation from renewables has only grown by 0.8 per cent per year since 1990. The participation of renewables in the total supply of electricity has decreased in absolute terms in many regions of the OECD since the late 1990s, for example in North America, particularly in the US. The European Union, on the other hand, has had a continuous growth since 1990, thanks to supportive policies, not least those related to urban waste handling.

Biomass only corresponds to 1 per cent of the world electricity generation. More specifically, electricity generation from solid biomass has shown an average increase of 2.7 per cent per year and some 20 TWh have been added to the supply base of OECD countries since 1990, denoting a slight increase in the share of biomass for electricity generation in OECD countries. In fact, renewable municipal waste and biogas are becoming increasingly important in OECD countries. Though both are still at an initial stage, we should expect significant growth in these segments in the years to come. Heat production from biomass has also increased substantially, both in heat plants only and in CHPs, but available data series do not allow further inference.

Biomass currently supplies 3.5 per cent of the energy in the EU, which is equivalent to 45 million toe. However, the interest for bioenergy has increased rapidly both among members and candidate countries. Some EU countries have had outstanding performance in their national biomass programs, for example, the Netherlands, United Kingdom and Denmark, all of which started from very low levels in the early 1990s. Also countries previously outside the EU such as the Czech Republic and Hungary have been investing in bioenergy (IEA, 2003).

In the past few years, the EU has developed common guidelines and energy directives, which are expected to have a significant impact in the coming years, not least on bioenergy use (see also Bauen, Chapter 2). Provided the efforts being made to promote bioenergy succeed, the amount could increase from 45 to 130 million toe in the region by 2010–15. Bioenergy provides a great opportunity to address problems other than energy in the EU, such as decreasing populations in rural areas, employment in peripheral regions, and restructuring of agricultural policies including new uses for idle croplands and reduction of subsidies. A recent Europe-wide study indicates that as many as 900 000 jobs could be created by 2020 from investments in renewables of which 500 000 are in agriculture to produce biofuels (ALTENER, 2001).

In developing countries, electricity generation from renewables has grown by some 3 per cent since 1990, following a parallel track with the increase in electricity demand in these countries at large. Thus the growth of renewables in electricity generation is larger in developing countries than in OECD countries. Certainly, most additions in the developing world come from hydropower, and only few countries are exploring other renewable sources systematically. Indeed, hydropower remains a major renewable option where potential is available. The conventional view favoring centralized energy generation may lead to large-scale projects, heavy financial burden on poor economies and negative environmental impacts. Yet many developing countries do have programs for small hydroplants.

Thus the truth about renewables is that there is a positive trend which may look impressive in relative terms but which is slow in absolute terms. This means that non-renewables not only remain very strong but are still mainstream. When it comes to biomass, the development has been slow in comparison with the new renewables. It can certainly be accelerated, bringing ancillary advantages to many countries, for example, rural development. To facilitate this process, there is a need for models that allow an effective and rapid assessment of local biomass potentials, while also providing guidelines to support project design and implementation. Certainly, there is no reason for allowing a very rapid move towards fossil fuels in developing countries where a significant untapped biomass potential exits. Reversing that trend is a major global challenge, and the introduction of bioenergy options definitely provides part of the solution.

1.4. THE TURNING POINT

We referred earlier as to what may be a turning point in bioenergy utilization. This idea needs perhaps to be further developed and motivated. During the industrialization period, started in England in the middle of the eighteenth century, fossil fuels gained increasing importance, offering the scale, efficiency and reliability needed to change production systems radically. The supremacy of fossil fuels was reinforced with the advent of the automobile and the choice of oil as the source of liquid fuels to move those engines. This process continued with full speed until three decades ago when oil-producing countries, in a concerted action, forced oil prices up to appropriate larger rents for a resource that the world economy so heavily relied upon.

After the oil price shocks, intense efforts were made to deploy new energy technologies based on resources other than oil, and to improve the efficiency of energy generation, distribution and consumption. Parallel to these efforts, however, a very significant amount of research continued being made on fossil-fuel-related technologies and nuclear power. As a result, while renewable technologies were indeed developed, the relative position of fossil-related technologies was constantly improved both on the supply as well as on the demand side. In addition, most of the non-fossil energy generation capacity added in the last few decades comes from nuclear power, an area that also received significant attention of governments and researchers.

Nevertheless, the balance of efforts made in the last few decades includes a portfolio of renewables, in parallel with a significant decrease in the energy intensity of many segments of the economy, and a trend of decarbonization mainly due to the shift from coal and oil towards natural gas and increasing use of nuclear power (Silveira, 2001). Whether positive trends will persist and be further improved depends on what efforts are made next. For example, recent studies reveal that increasing amount of investments is no guarantee for improvements in energy intensity, as expanding industries can, in fact, become more energy intensive (Miketa, 2001). We are also used to the thought that the energy intensity of developing countries shall increase as a result of industrialization and modernization. However, if we consider the increasing utilization of combustible renewables and wastes, the energy intensity may have decreased in some developing countries in the past years. Constraints in the utilization of combustible renewables and wastes may be forcing a higher utilization of fossil fuels in developing economies than would otherwise be necessary (Sun, 2003).

When it comes to transport, the sector remains trapped in the oil solution after three decades of research and constant improvements. More recently, the strong dependency of the transport sector on oil resources has received increasing attention due to issues of security, potential oil scarcity in a rather near future, and the climate change agenda (see also Silveira, 2001). In the European Union, for example, security of supply and climate change are two major driving forces to the introduction of renewables. Liquid biofuels provide immediate opportunities to reduce fossil fuel dependency in the transport sector, taking advantage of existing distribution chains for fossil fuels. A major preoccupation is the formation of markets for alternative transport fuels and technologies. However, significant initial steps have been taken recently at the EU level which may have important consequences in the development of markets for liquid biofuels.

But what is actually the turning point that we are referring to? After all, the figures do not indicate any spectacular change in favor of bioenergy. The use of bioenergy is actually growing slower in many cases when compared with other renewable options. Recent trends do not, at first, seem to relate to ambitious future scenarios and identified possibilities for bioenergy options. In fact, the turning point can only be understood as a convergence of factors and tendencies that are likely to favor bioenergy use. Some of these factors are general for all renewables, others are specifically related to bioenergy options. The most important factors are:

- The global climate agenda, which requires a shift from fossil fuels to renewables as a means to reduce greenhouse gas emissions and mitigate global climate change;
- Increasing awareness and understanding of the local impacts of fossil fuel utilization on environment and health (e.g. acid rain, respiratory diseases) and intensified search for sustainable alternatives;
- Decreasing policy support for fossil fuels and gradual reduction of subsidies for nonrenewable energy sources;
- A shift from centralized energy planning due to the privatization of electricity and heat markets, favoring local alternatives and decentralized solutions;

- Awareness of the potential of bioenergy options to foster regional development (e.g. through the creation of jobs);
- Enhancing policy support framework for renewables, including bioenergy, in many countries and regions (e.g. various EU directives);
- Better understanding of the potential integration of bioenergy solutions with established industrial processes leading to economic and environmental benefits (e.g. forest industry);
- Integration of bioenergy options with established energy systems for heat, power and transport (e.g. cofiring, ethanol additives);
- Critical mass of examples of good performance of bioenergy systems including biofuel production, heat and power generation, and demand-side technologies in various countries under different conditions;
- Variety of scales, raw material sources and technologies that can be used for the implementation of bioenergy systems depending on local conditions for raw material production, existing demand for energy services and future potential for expansion;
- Improving conditions for new entries and competition as biofuel markets evolve and the commercial attractiveness of bioenergy options is improved, while risks are reduced;
- Readily available conventional solutions and promising development of new technologies for bioenergy generation, and biofuel production and utilization.

These factors are processes which, when combined in different institutional and regional contexts, have varied impacts and effects. They contrast with the set of conditions that allowed the extraordinary economic development in the past decades and which included cheap energy provided by fossil fuels, lack of environmental concerns and centralized energy planning (see also Schipper et al., 1992). Energy was particularly cheap because there was little preoccupation to internalize the various costs associated with its extraction, transport and use, let alone with the sustainability of environmental and socioeconomic systems.

The present conditions are quite different. In particular, it is most likely that both households and industry will experience significant increase in energy costs in the coming years, if international efforts prove fruitful in moving the environmental agenda forward. In the medium and long term, it is possible to improve the overall efficiency of production systems through dematerialization, new industrial organizational patterns and new forms of land use planning. Due to its potential integration into various production segments and its role in social and environmental sustainability, bioenergy can become an important element in the process of shifting energy systems. The convergence of factors required to reach a turning point has already been reached. Work lies ahead.

1.5. TAKING THE LEAP TOWARDS BIOENERGY

In the past decade, the number of countries exploring biomass opportunities for the provision of energy services has increased rapidly. This has contributed to make biomass, in the form of solid and liquid fuels, an attractive and promising option among available renewable energy sources. This includes solid biomass and waste, which consists of firewood, charcoal, energy crops, and forest and agricultural residues for the production of heat and power, as well as short crops for the production of liquid fuels such as ethanol and biodiesel. Also the increasing attention to urban waste has contributed in drawing attention to bioenergy options. What is in place is a result of combined top down and bottom up initiatives. However, nothing seems more powerful at present than the increasing awareness about biomass potentials resulting from successful experiences in both industrialized and developing countries.

We still need a much more forceful move towards renewables if we are to promote our energy systems to a qualitative leap. In this context, bioenergy offers attractive alternatives which are only partially being explored. The enhancement of bioenergy utilization has to count on modern and efficient technologies, which should be deployed on a commercial basis in order to guarantee energy services of high quality. Commercial options are sorted within competitive markets. But how can we talk about competition between bioenergy and other alternatives when choices are not on the table at a fair playing field?

Recent studies indicate that biomass technologies can be competitive with fossil fuel alternatives. One particular advantage of bioenergy is that it can be organized at small scales, from 1 to 100 MW, thus allowing a slow modular increment in energy supply, avoiding stranded investments, and minimizing risks. At a time of restructuring of the electricity sector, these are essential advantages, as economies of scale may not be easily realized in volatile markets. In addition, risk aversion and high demand for faster returns by stakeholders will tend to favor smaller projects and a gradual change in the configuration of the electricity infrastructure (Patterson, 1999). The solar economy, which includes bioenergy, favors small-scale and decentralized solutions with local distribution, which differ significantly from the centralized and large-scale configuration of existing energy systems (Wicker et al., 2002; Scheer, 1999).

Bioenergy is not a generalized solution for all countries and regions. The dimension of the regional potential for bioenergy needs to be seen in the context of competing uses for resources demanded for the production of biomass. Where land resources are scarce, energy forests may compete with other land uses and lead to negative impacts on food production. However, there are many countries in the world where this is not the case. Many developing countries such as Brazil, Thailand,

Indonesia and Nigeria have large amounts of biomass potential from different sources and are good candidates for bioenergy technologies.

In Europe, the restructuring of agriculture is releasing land, which can be claimed for biomass production aimed at energy generation. If biomass is to become a major source of energy in Europe as a whole, the potential needs to be assessed in terms of the overall environmental and socioeconomic implications vis a vis other alternatives. The possibility of increasing supply security through a broader use of bioenergy needs is to be more seriously considered. A more significant increase in the share of biofuels cannot be attained through isolated national initiatives but will require coordinated action, not least to facilitate the formation of biofuel markets (European Commission, 2000).

There is a long way to go before bioenergy becomes a mainstream energy alternative. In particular, there are significant market barriers to be overcome, which can only be achieved through close coordination among the various sectors that need to be involved in bioenergy initiatives. This book discusses some of the opportunities that are already at hand to harness the bioenergy potential and some of the progress that has been achieved in different contexts.

The turning point should rather be understood as a perception among experts, policy makers and industries that a wide window of opportunity has been opened, which should be used to realize the global bioenergy potential. In many cases, the leap is more political than economic given, for example, that the removal of subsidies from nonrenewable alternatives is a necessary step in the process. In many regions, political coordination of efforts is a necessary initial step to establish bioenergy markets. In any case, the leap towards a broader utilization of bioenergy is now more psychological than technological.

REFERENCES

ALTENER (2001) The impact of renewable on employment and economic growth, EU report.

EEA – European Environment Agency (2002) Energy and Environment in the European Union, Copenhagen.

EU (2001) National Energy Policy Overview, available at http://energytrends.pnl.gov/eu/eu004.htm on June 23, 2004.

European Commission (2000) Green Paper Towards a European Strategy for the Security of Energy Supply, available at http://europa.eu.int/comm/energy_transport/livrevert/final/report_en.pdf on June 23, 2004.

Hall, D. & Rosillo-Calle, F. (1993) *Why Biomass Matters: Energy and the Environment in Energy in Africa*, Ponte Press, Bochum, Germany.

IEA (1997) Biomass Energy: Key Issues and Priority Needs, Paris.

IEA (2002) Renewables in Global Energy Supply, an IEA Fact Sheet, available at http://library.iea.org/dbtw-wpd/textbase/papers/2002/leaflet.pdf on June 23, 2004.

IEA (2003) Renewables Information, Paris.

IEA (2004) *Biofuels for Transport – an International Perspective*, Paris.

Kaijser, A. (2001) From Tile Stoves to Nuclear Plants – the history of Swedish energy systems, in *Building Sustainable Energy Systems – Swedish Experiences*, Ed. Silveira, S., Swedish Energy Agency, Eskilstuna, Sweden.

Meyer, N.I. (2003) European Schemes to promote renewables in liberalized markets in *Energy Policy*, **Vol. 31**(7), Elsevier, pp 665–67.

Miketa, A. (2001) Analysis of energy intensity developments in manufacturing sectors in industrialized and developing countries in *Energy Policy*, **Vol. 29**(10), Elsevier, pp 769–775.

Nakicenovic, N. et al. (1998) *Global Energy Perspectives*, Cambridge University Press, Cambridge.

Scheer, H. (1999) *The Solar Economy*, Earthscan, London, UK.

Schipper, L. & Meyers, S. (1992) *Energy Efficiency and Human Activity: Past Trends, Future Prospects*, Cambridge University Press, New York.

Silveira, S. (2001) Tranformation of the energy sector, in *Building Sustainable Energy Systems – Swedish Experiences*, Ed. Silveira, S., Swedish Energy Agency, Eskistuna, Sweden.

Sun, J.W. (2003) Three types of decline in energy intensity – an explanation for the decline of energy intensity in some developing countries in *Energy Policy*, **Vol. 31**(6), Elsevier, pp 519–526.

WEC (2001) *Living in One World*, London, Great Britain.

Wicker et al. (2002) *Small is Profitable*, Earthscan, UK.

Woods, J. & Hall, D.O. (1994) Bioenergy for Development – Technical and Environmental Dimensions, FAO Environment and Energy Paper 13, Rome.

Chapter 2

Biomass in Europe

Ausilio Bauen

2.1. IS BIOMASS IMPORTANT TO EUROPE?

Energy, environment, agricultural and forestry-based drivers are contributing to a rediscovery of bioenergy in industrialized nations with access to biomass resources. In fact, bioenergy offers the possibility to harness a domestic, rural-based, low-carbon and sustainable energy source in both industrialized and developing countries. Currently, commercial and noncommercial uses of biomass represent about 13.5 per cent of the world's primary energy consumption (see also Figure 1.1).

In the European Union (EU), bioenergy comprises some 3.5 per cent of the total primary energy mix. Figure 2.1 shows the primary energy consumption in the European Union, including details of renewable energy sources. Notably, biomass is the largest renewable energy source in the European Union. The biomass resources commonly used in the EU are fuelwood, wood residues from the wood-processing industry, used wood products (e.g. demolition wood), and also straw in some countries. Various modern technologies are being applied.

Bioenergy is intrinsically linked to energy, environment, agriculture and forestry issues. As such, it receives consideration within international and national renewable energy, as well as environment, agriculture and forestry policy agendas. Unfortunately, there is a lack of integration across these policy agendas, which hinders the understanding of constraints affecting bioenergy, and the convergence of incentives to promote it, ultimately delaying its development.

Two fundamental questions related to the development of bioenergy options are: (i) what biomass conversion technologies and end-uses will present the most favorable economic and environmental options in the future energy mix; and (ii) what amount of biomass resources will these options require? Options range from heat and power production to liquid-fuel substitutes, but opinions vary widely on their potential contribution to future energy mixes and with regard to the appropriate resources, technologies and scales that are to be applied.

Questions as to which short-term bioenergy options are practical, and where the opportunities lie for establishing markets for biofuels and bioenergy technologies

Bioenergy – Realizing the Potential

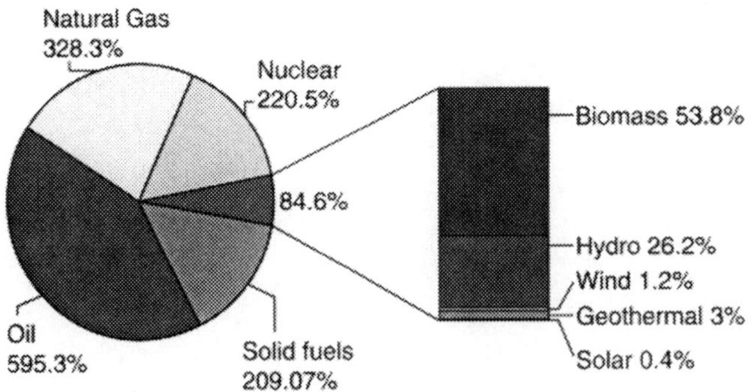

Figure 2.1. Total primary energy mix in Europe (EU 15). Source: European Commission (2001a).

in the near future are key to further development in this area. Also, it is important to verify how short-term developments fit with the potential long-term role envisaged for biomass in the energy mix. Interesting short-term markets for bioenergy appear to exist for cofiring with coal, district and small-scale heating, combined heat and power and blending of biofuels with petroleum transport fuels. Long-term options could be biomass use for heat and power generation in integrated gasification combined cycle plants and for the production of new fuels such as hydrogen.

This chapter briefly discusses the bioenergy potential in Europe and some of the energy, environment and agriculture cross-cutting issues that are relevant in the definition of coordinated action for bioenergy in the European context. Climate change issues and long-term policies for renewables are likely to have a significant impact on the development of bioenergy and these issues are, therefore, particularly addressed.

2.2. BIOMASS RESOURCES AND CONVERSION TECHNOLOGIES

Biomass is available in a variety of forms and is generally classified according to its source (animal or plant) or according to its phase (solid, liquid or gaseous). Generally, bioenergy can be derived from sources such as forests and energy crop plantations, residues from primary biomass production, and by-products and wastes from various industrial processes.

Forests, woodlands, short rotation forestry and other arboricultural activities (for example, park maintenance) are a source of wood fuel. Fuel can also be obtained from energy crop plantations using species such as willow, eucalyptus, sugarcane, miscanthus, energy grain, hemp, oilseed rape, sunflower and sugar beet. Residues represent another possible source of fuel. This includes residues from food and industrial crop production (for example, cereals, sugarcane, tea, coffee, rubber trees, oil and coconut palms) and residues from forestry activities (for example, from stem wood production). By-products and wastes may also originate from sawmill waste, manure, sewage sludge, abattoir waste and municipal solid waste. Generally, these are sources of low-cost fuel.

Biomass and waste needing disposal can be burned directly or converted to intermediate solid, liquid or gaseous fuels to produce heat, electricity and transport fuels. A number of biomass conversion technologies are currently commercially available. In addition, there is a potential for technological advances and commercialization of more efficient technologies for production of electricity and transport fuels in a rather near future. Table 2.1 shows a range of biomass technology options and corresponding end-uses, indicating also the status of these technologies.

There are significant differences among European countries when it comes to the exploitation of biomass resources. The bulk of biomass being used consists of fuelwood for domestic heating. The use of biomass for district heating is substantial in a few countries such as Austria, Finland and Sweden, mainly fed by fuelwood and wood residues from the forestry and wood-processing industry. In Denmark, straw is used to some extent. In comparison, the use of biomass in industry and for power generation is modest. In some countries, such as Sweden, electricity is generated in combined heat and power plants connected to district heating. In addition, biofuels in transport applications represent a small fraction of the bioenergy use in countries such as Austria, France, Germany, Italy and Sweden.

A large biomass potential remains unexploited in Europe, for example in the form of residues from woodland management measures, agricultural residues, organic waste from industry and households and energy crops (see also Table 11.1). The total biomass potential is estimated at 6759.2 PJ (161.4 Mtoe), with the largest contributions coming from woody residues (i.e. wood residues from stem wood production, thinning from managed forests, and wood waste from the wood products industry and arboricultural activities) and from a variety of annual and perennial energy crops (Bauen and Kaltschmitt, 2001).

2.3. THE ROLE OF BIOMASS IN CLIMATE CHANGE MITIGATION

The need for reductions of greenhouse gas emissions may provide a significant incentive to further develop bioenergy. Biomass can act as a carbon sink and as a substitute for fossil fuels. Its role as a means of reducing CO_2 in the atmosphere is recognized in the Kyoto Protocol in articles 3.3 and 3.4. The IPCC (1995) estimates that between 60 and 87 GtC could be stored in forests between 1990 and 2050, corresponding to about 12–15 per cent of projected fossil fuel emissions, and without regard to carbon storage in biofuel plantations in currently unforested land.

Table 2.1. Biomass technology options, corresponding end-uses and status

Conversion technology	Resource type	Examples of fuels	Product	End-use	Technology status
Combustion	Mainly solid biomass	Wood logs, chips and pellets, solid waste, chicken litter	Heat	Heat Electricity (steam turbine)	Commercial
Gasification	Mainly solid biomass	Wood chips and pellets, solid waste	Syngas	Heat (boiler), Electricity (engine, gas turbine, fuel cell, combined cycles), Transport fuels (e.g. diesel, methanol, hydrogen)	Demonstration/ Early commercial
Pyrolysis	Mainly solid biomass	Wood chips and pellets, solid waste	Pyrolysis oil + by-products	Heat (boiler), Electricity (engine)	Demonstration/ Early commercial
Pressing/ Esterification	Oleagenous crops	Oilseed rape	Biodiesel	Heat (boiler), Electricity (engine), Transport fuel	Commercial
Fermentation/ Hydrolysis	Sugar/starch/ lignocellulose	Sugarbeet, corn, fibrous and woody biomass	Ethanol	Transport fuel	Commercial/ early demonstration
Anaerobic digestion	Wet biomass	Manure, sewage sludge	Biogas + by-products	Heat (boiler), Electricity (engine, gas turbine, fuel cell)	Commercial

While the establishment of forest-based carbon sinks may have an important role, they are by no means the solution to climate change. They also remain contentious, a principal concern being related to the permanency of the sink. Hence, there is a view that biomass sinks should be associated with a multifunctional role for biomass, be it for the production of bioenergy or raw materials for other purposes (Schlamadinger et al., 2001; Read, 1997).

The advantage of using sustainably grown biomass for energy is that it ensures emissions reductions through the substitution of fossil fuels and is not constrained by the saturation limits of managed biomass carbon sinks. Bioenergy for fossil fuel substitution may be complemented with significant carbon sequestration in litter and soils, depending on land-use changes. The levels of carbon substitution and sequestration will depend on the plant species grown and associated management practices, as well as on soil types. Land use and management directed at using biomass for fossil fuel and other raw material substitution could reduce concerns over the temporary nature of land use changes for carbon mitigation as it would be linked to a traded commodity in the form of biomass materials. Associated carbon sinks could also be more secure.

Changes in land use and land management practices associated with energy crops as well as biofuel chain logistics affect the carbon cycle. Consequently, energy crops are not necessarily carbon neutral. The magnitude of carbon released or stored both above and below the ground through the introduction of energy crops may significantly affect the carbon balance of biofuel cycles. This needs to be considered in determining the carbon credits that can be attributed to them. Generally, the introduction of herbaceous and woody perennials on agricultural land or degraded land will lead to an increase in soil carbon. However, many factors, including those external to the land use and management practices, such as local climate, will affect the soil carbon balance and may lead to uncertainties in its assessment. Concerns still remain over the permanence of the carbon sinks.

Following from the Kyoto Protocol, the EU target is a reduction of 8 per cent of greenhouse gas emissions by 2012. Biomass already contributes to avoided CO_2 emissions by supplying part of the energy demand in the European Union, which would otherwise be mainly met with fossil fuels. Avoided CO_2 emissions associated with current biomass use are estimated at 2–9 per cent of the 1998 energy-related CO_2 emissions in the EU (Bauen and Kaltschmitt, 1999).

Increased utilization of biomass could make a substantial additional contribution to reduce CO_2 emissions and meet the Kyoto Protocol targets. Based on potential estimates and assumptions on its use for heat and power purposes only, it is believed that biomass could reduce 1998 CO_2 emissions by between 6 and 26 per cent (Bauen and Kaltschmitt, 2001). The consideration of carbon sinks could add further reductions of CO_2 emissions as a result of biomass utilization (Schlamadinger et al., 2001).

The EU climate policy emphasizes nitrogenous emissions from agricultural activities. Perennial grasses and woody crops have a lower nitrogen fertilization demand and higher nitrogen use efficiency compared to annual crops (including annual biofuel crops), leading to lower nitrogen losses (nitrogen leaching and gaseous nitrogen emissions, mainly N_2O). Hence, the choice of energy crops will affect greenhouse gas emissions from the agricultural sector and may influence climate change mitigation actions. Nitrogen losses are subject to uncertainties, as they will be affected by soil type.

How such aspects may translate into policy actions that can be integrated with other agricultural, energy and environmental policies deserves further consideration. Although carbon sinks remain contentious and the extent to which they should contribute to the Kyoto commitments unsure, land management for fossil fuel substitution is to some degree likely to be a key issue in meeting stringent greenhouse gas emissions targets. An important issue yet to be addressed is if and how carbon sequestration associated with land management for fossil fuel substitution should be considered.

2.4. THE EU ENERGY AND AGRICULTURE POLICIES

The main goals of the European Union in the energy sector, including energy use in transport, are to meet the Kyoto Protocol objectives, double the share of renewable energy supply by 2010, improve energy efficiency, security and diversity of supply, enhance competitiveness of European industries and create jobs. So far, the EU has developed recommendations on energy policy goals but has not had a legal basis (*competence*) for energy policy per se – this has been the responsibility of Member States. This situation is about to be changed as energy shall now be recognized as an area of shared responsibility between the EU and Member States individually.

Bioenergy is projected to become a major contributor to the EU's future primary energy mix. It already contributes about 60 per cent of the renewable energy share in the European Union and is believed to be the renewable energy with the largest growth potential. The White Paper on Renewable Energy (European Commission, 1997) estimates the contribution of biomass and waste in 2010 at 135 million toe, still representing about 60 per cent of the total primary renewable energy in the region.

A significant increase in biomass utilization will be needed to achieve this goal. This includes an additional 15 million toe from biogas exploitation, 30 million toe from agricultural and forestry residues, and 45 million toe from energy crops, requiring 10 million hectares out of the 77 million hectares of agricultural land in the EU. For comparison, it can be mentioned that EU-15 set-aside land for agriculture is about 9.5 Mha, which grew to 30 Mha with EU enlargement. It is also estimated that

18 million toe of biomass from energy crops will be used for the production of liquid biofuels (biodiesel and ethanol) in 2010. As a matter of fact, a recent directive on the promotion of biofuels for transport aims at facilitating their introduction (European Commission, 2003). The objective is to replace 2 per cent of fossil fuels in transport with biofuels by 2005, and 5.75 per cent by 2010.

Meeting the EU renewable energy targets for biomass obviously requires a significant amount of land and other resources. Over recent years, set-aside agricultural land has been of the order of 20 per cent, and between 10 and 15 per cent of this land has been devoted to nonfood crops. However, reform of the Common Agricultural Policy (CAP) will lead to reductions in the share of land set aside, though in absolute terms this is compensated by the expansion of the European Union. Enlargement of the EU can have significant implications for bioenergy, not least due to the implied increase in agricultural land. In addition, agriculture in accession countries is likely to witness significant improvements in efficiency as a result of increasing economic pressure and competition, thus farmers may welcome the diversification to nonfood energy crops. Oilseed rape covers about 80 per cent of the set-aside land devoted to nonfood crops in the EU, though its cover has decreased considerably in the last few years and the area covered by sugar beet (as a nonfood crop) and short rotation coppice (SRC) has been growing.

In 1998, about 438 000 ha of set-aside land were dedicated to crops for liquid biofuels production (340 000 ha of rapeseed, 68 000 ha of sunflower seed, 18 000 ha of cereals and 12 000 ha of sugar beet) and about 20 000 ha of SRC, mainly for heat generation via direct combustion. Fuelwood production continues to rise in the EU and better forest management and the establishment of new forests could contribute significantly more wood fuel, as is believed to be the case in Sweden (see also Ling and Silveira, Chapter 3). Large agricultural holdings ($\geq$100 ha) contribute more than two-thirds of the total land used for nonfood products, and more than half of the total land under incentive schemes. Nevertheless, land used to produce agricultural raw material for nonfood purposes covers less than 1 per cent of the total cultivated area of these holdings.

The Common Agricultural Policy (CAP) reforms have two goals: the first is an increasing market orientation of the sector, and the second is the reinforcement of structural, environmental and rural development aspects of sustainable agriculture (European Commission, 2002). Aspects that could favor bioenergy, such as the multifunctionality of agriculture, are amongst the principles driving agricultural policy in the EU today. Although there are no specific nonfood policies, a number of measures related to agri-environment and structural measures in particular, provide opportunities for the development of nonfood crops.

The role of the agricultural sector in reducing greenhouse gas emissions is gaining relevance. Agriculture contributed towards 11 per cent of the EU's greenhouse gas emissions in 1990, mainly with CH_4 and N_2O emissions. The agricultural sector has been included in the European Climate Change Programme adopted by the Commission in March 2000 as an area for common action. The working group on agriculture recognizes not only the importance of reducing greenhouse gas emissions associated with agricultural activities, but also the importance of agriculture as a carbon sink and provider of renewable raw materials to the energy and industrial sectors (European Commission, 2001b).

To increase biomass contribution as a renewable energy source, reduce greenhouse gas emissions and promote sustainable rural development, there needs to be a better understanding of the economic and environmental implications of different biofuel chains, their role in an evolving energy sector, and in regional development. The promotion of bioenergy should then be driven by policies aimed at seizing the environmental and rural development benefits of bioenergy in an integrated manner.

2.5. EXAMPLES OF COUNTRY POLICIES WITHIN THE EU

Bioenergy and related policies in the UK

The UK has ambitious greenhouse gas emissions reduction targets, that is, 20 per cent reduction by 2010 compared to 1990 levels. When it comes to the role of renewable energy, the target is 10 per cent renewable energy generation by 2010. The government has been introducing a number of policies aimed at fulfilling these targets, such as the Renewables Obligation and Green Fuels Challenge.

The government's renewable energy target is ambitious. Approximately one-third of the renewable energy is expected to come from biomass, which may require up to 125 000 ha of energy crops for power generation. The Non-Fossil Fuel Obligation has contributed to stimulating the market for renewables somewhat, particularly for wind power, but there has generally been little incentive for developing renewable energy sources. The main policy pushing for renewable electricity is expected to be the Renewables Obligation, which forces electricity suppliers to provide a fraction of their electricity from renewables.

However, the proposed buy-out price of 3.5 pence/kWh (5.5 €cent/kWh) is not likely to promote significant investments in bioenergy schemes other than those using low-cost biomass waste as fuel, unless other policies are adopted. Recently, the government has allocated an additional £100 million (€160 million) to the development of renewable energy, a significant part of which shall be destined to heat and power from biomass.

Under the Green Fuels Challenge, biodiesel will qualify for a 20 pence per liter fuel duty rebate from 2002. Unfortunately, the tax reduction proposed is unlikely to make biodiesel competitive with diesel, except if produced from recycled vegetable oils. Much discussion has surrounded this tax reduction issue, with biodiesel supporters asking for a significantly higher reduction and government claiming that it would not be justifiable on environmental grounds. From January 2005, bioethanol will also qualify for a 20 pence per liter fuel duty rebate. The UK government is now in the process of setting UK targets under the EU biofuels directive, and additional mechanisms may be required to assist in meeting the targets.

The UK Department of Environment, Food and Rural Affairs (DEFRA) has allocated £30 million for the introduction of energy crops (SRC and Miscanthus). The area-based incentives will cover part of the set-up costs and should lead to the establishment of approximately 6000 ha of energy crops. The SRC also have access to support from the Forestry Commission's Woodland Grant Scheme. The production of oilseed rape for nonfood purposes benefits from subsidy payments under the current EU CAP set-aside policy which is aimed at reducing arable land area dedicated to food production. There is no other specific incentive for its cultivation for biodiesel production in the UK.

A UK Emissions Trading Scheme for greenhouse gases (GHG) has been in place since 2002 and has acted as a precursor to the EU-wide trading scheme to begin in 2005. The GHG trading schemes could support biomass for energy, but how the carbon substitution and carbon sink benefits from biomass energy will be treated by these schemes remains unclear. There are also opportunities for funding demonstration schemes through funds derived from the Climate Change Levy.

There appears to be an increasing, albeit limited, number of initiatives dedicated to the development of bioenergy in the context of energy, agriculture and climate change policies in the UK. However, there is no clear integrated strategy for the promotion of bioenergy based on short- to long-term considerations. Such considerations should include analyses of environmental and socioeconomic impacts of different pathways on both the energy and the agricultural sector, as well as on regional development.

Bioenergy and related policies in Italy

Bioenergy use in Italy is relatively low at about 3.5 Mtoe/year. Biomass is used for single-house heating and district heating schemes, using forestry residues, mainly in northern Italy. It is also used in CHPs, based on agricultural and food industry waste, and for biodiesel production (approximately 100 000 t/year). Italy's heavy reliance on energy imports especially fossil fuels for electricity generation and environmental considerations are driving forces favoring bioenergy. Also the availability of significant biomass resources from agricultural and agro-industrial wastes, energy crops on set-aside agricultural land and wood fuel from improved forest management and reforestation are contributing to raise the interest in bioenergy.

A number of recent government policies related to energy, environment and agriculture are indications of this interest. The Italian government has approved a White Paper on renewable energy and national guidelines for the reduction of greenhouse gases. The White Paper indicates a target of 8–10 Mtoe for primary energy from biomass (24 Mtoe for all renewables). This policy paper is complemented by a Biomass Implementation Programme based on the action plan National Programme for Renewable Energy from Biomass, designed by the Ministry of Agriculture. Fiscal incentives are directed at biodiesel for transport, by which 300 000 t/year are exempted from taxes over a period of three years. No quotas are placed on biodiesel for heating purposes, which is also tax exempt. The Italian government is now in the process of setting targets for Italy under the EU biofuels directive.

However, important barriers persist, such as the lack of a nonfood crop policy, organizational barriers in terms of concerted actions on the part of stakeholders in the bioenergy chain, and possibly some public opposition to biomass schemes. Progress in joined-up thinking needs to consider more closely the benefits and risks of bioenergy, innovation in the bioenergy chains, and organizational and institutional barriers.

Bioenergy and related policies in the Netherlands

The contribution of renewables to the Dutch energy mix is relatively low, at around 1 per cent. However, the country has taken a fairly proactive stance in energy and environment issues and aims at a 5 per cent renewable energy share in 2010. Bioenergy is expected to provide about half of the renewable target, rising from about 13 to about 70 PJ/year (excluding waste incineration). The Dutch government is now in the process of setting targets for the Netherlands under the EU biofuels directive.

A number of policies are being directed mainly at the energy sector. These consist of fiscal instruments and green funds and agreements in various sectors of the bioenergy chain, such as commitments on the part of biomass suppliers, generators (e.g. cocombustion in coal plants) and end-users (e.g. industry and municipalities). Demand and willingness to pay for green electricity is also expected to act as a driving force. The main barriers to bioenergy remain the availability of biomass, the profitability of bioenergy schemes and the integration and continuity of relevant policies.

2.6. CONCLUDING REMARKS

Biomass has the potential to become a major contributor to the European primary energy mix for the supply of modern energy services. The extent to which bioenergy uptake will occur, and its rate of uptake, will depend on resource availability, economic and environmental constraints, as well as policy measures resulting from drivers such as climate change and willingness to enhance energy supply independence.

Biomass may be used to provide a number of energy vectors through various fuel chains. Most of these will present benefits in terms of displacing and saving nonrenewable energy sources, reducing greenhouse gas emissions and providing income diversification to farmers. However, the economic and environmental characteristics of the fuel chains and their ability to supply the energy vectors of the future may vary considerably. Biomass can become an important renewable energy source in industrialized countries only if it is able to supply the energy vectors demanded by modern energy services based on environmentally and economically sound fuel chains.

Thus bioenergy incentives must account for the environmental characteristics of the fuel chain i.e. from the production of the fuel to the energy service provided. A variety of market-based mechanisms can be applied at different stages of the fuel chain to stimulate development. In the case of energy crops, mechanisms need to be devised in greater synergy among energy, agriculture and environmental policies to encourage farmers to grow biomass resources in a sustainable manner.

If current energy market structures and policies are maintained, renewable energy penetration, including biomass, is likely to remain low. Piecemeal policies directed at bioenergy are being introduced in a number of EU countries as exemplified here. However, apart from countries which already have a significant biomass resource base, the uptake of biomass energy has been slow. Without the contribution of biomass, it will be difficult to meet the carbon emission reductions envisaged by the Kyoto Protocol, let alone further reductions likely to be required in the post-Kyoto period. Biomass can play a substantial role in greenhouse gas reductions, and it is important to enhance understanding of carbon stock and fossil fuel substitution dynamics. Mechanisms are also needed to provide incentives for fossil-fuel substitution and for the development of sustainable long-term carbon sinks.

Clearly, much needs yet to be done in identifying and implementing viable bioenergy pathways that could contribute to a low-carbon future. Short- to long-term strategies need to be defined and enabling policies, designed and implemented. In particular, there is an urgent need for policy integration to make different bioenergy drivers converge, catalyzing economic and environment beneficial uses.

Actions delivering win–win–win situations across the agriculture, energy and the environment need to be further explored.

REFERENCES

Bauen, A. & Kaltschmitt, M. (1999) Contribution of biomass toward CO_2 reduction in Europe, *Proceedings of the 4th Biomass Conference of the Americas*, August 29–September 2, Oakland, California, Elsevier Science.

Bauen, A. & Kaltschmitt, M. (2001) Current use and potential of solid biomass in developing countries and their implications for CO_2 emissions, *Proceedings of the 1st World Conference and Exhibition on Biomass for Energy and Industry*, 5–9 June 2000, Seville, Elsevier Science.

European Commission (1997) Energy for the Future: Renewable Sources of Energy – White Paper for a Community Strategy and Action Plan, COM(97)599 final, Brussels.

European Commission (2001a) Energy and Transport in Figures 2001, Directorate-General for Transport and Energy, Brussels.

European Commission (2001b) European Climate Change Programme – Long Report, Brussels.

European Commission (2002) Communication on Mid-term Review of the Common Agricultural Policy, COM(2002)394, Brussels.

European Commission (2003) Directive 2003/30/EC of The European Parliament and of The Council of 8 May 2003 on the Promotion of the Use of Biofuels or Other Renewable Fuels for Transport, Brussels.

IPCC – Intergovernmental Panel on Climate Change (1995) Second Assessment Report: Climate Change 1995, IPCC, Geneva.

Read, P. (1997) Food, fuel, fibre and faces to feed. Simulation studies of land use change for sustainable development in the 21st century in *Ecological Economics*, **Vol. 23**(2), Elsevier.

Schlamadinger, B., Grubb, M., Azar, C., Bauen, A. & Berndes, G. (2001) Carbon sinks and the CDM: could a bioenergy linkage offer a constructive compromise? in *Climate Policy*, **Vol. 1**(3), Earthscan.

Chapter 3
New Challenges for Bioenergy in Sweden

Erik Ling and Semida Silveira

3.1. BIOENERGY IN TRANSITION

The next ten years will be decisive in terms of turning biomass into a major modern and reliable energy supply source globally. The ongoing development of bioenergy technologies and the know-how and liberalization of energy markets, allied to increasing international trade with biofuels, and policies supporting emissions trading and green certificates are likely to create favorable conditions for a larger utilization of bioenergy.

Also for countries such as Sweden, where bioenergy already occupies a very significant place in the total energy mix, the conditions for the development and utilization of bioenergy are changing rapidly. In this chapter, we look briefly at what has been accomplished in Sweden in the past few decades, and discuss three major drives that both open opportunities and bring new challenges to the bioenergy segment. These drives need to be considered in the design and implementation of robust strategies for the sector. They refer to the internationalization of the bioenergy segment, integration of bioenergy systems with other production processes, and mainstreaming of bioenergy as a major energy source.

Further, we discuss more specific tasks that the Swedish bioenergy segment will have to deal with in the near future. These tasks are related to the energy supply source, integration within the forest industry, reduction of CO_2 emissions and development of a competitive bioenergy industry. The matrix format of our analysis is illustrated in Table 3.1. Though this discussion is focused on conditions observed in Sweden where bioenergy has evolved closely linked with forestry activities, we believe that it serves as a reference for other countries which are either contemplating the utilization of bioenergy or wanting to benefit from the formation of biofuel markets.

3.2. BIOMASS UTILIZATION IN SWEDEN

During the last few decades, the Swedish energy sector has undergone substantial changes. In short, nuclear power and biomass have become major energy sources,

Bioenergy – Realizing the Potential

Table 3.1. Drives affecting bioenergy development in Sweden, and major tasks to enhance opportunities

	Enhancement of the biomass supply source	Systems integration within the forest industry	Reduction of CO_2 emissions	Development of a competitive bioenergy industry
Internationalization of bioenergy segment	Increasing imports of biofuels/standardization	More effective forest/energy operations within the forestry sector	Emissions trading, CDM and JI as a means to promote bioenergy broadly	Critical mass needed/scale
	Transnational energy companies			Project clusters Standardization
Integration of bioenergy systems with other production processes	New dimensions of energy markets	Integrated forest and energy industries	Complex synergies, incentive structure and market signals	Industrial clusters
	Integration between energy sources and users			Systems solutions
Bioenergy as a mainstream alternative	Reduction of tax incentives/ more competition	Sustainable forest production	Establishment of institutions and market structures (e.g. finance)	Link bioenergy development to industrial development policy
		Need to review practices (e.g. return ashes) to guarantee system sustainability		

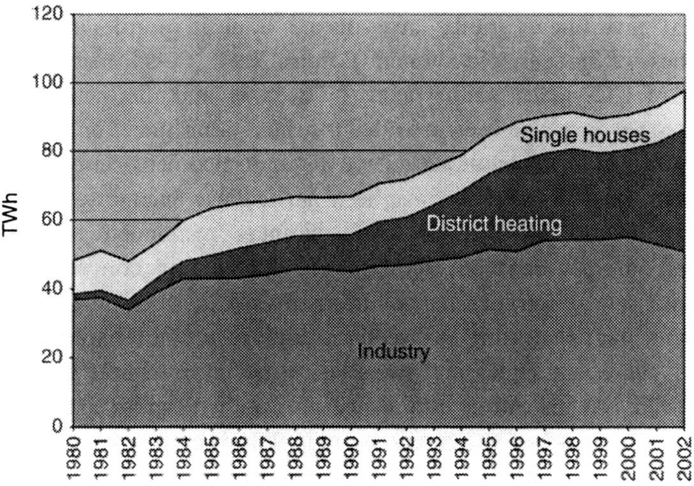

Figure 3.1. Biofuel use in Sweden in TWh, 1980–2002. Source: Swedish Energy Agency, 2003.

while the utilization of fossil fuels was radically reduced (Silveira, 2001). Total energy supply went from 457 TWh in 1970 to 616 TWh in 2002. However, in 1970, fossil fuels corresponded to 80 per cent of the total energy supply in the country compared with 38 per cent in 2002 (Swedish Energy Agency, 2003).

Sweden has reached remarkable progress in the utilization of renewable energy, not least in comparison with other OECD countries. In 2002, almost one-third of the total energy supply in the country came from renewable sources, mainly hydropower, biomass and windpower. As a result of focused attention on the national energy potential throughout the last few decades, biomass has attained a particular place in the Swedish energy system. Today, it corresponds to approximately one-fifth of the total energy used in the country, and is the number one energy source outside of the transport sector[1] (Swedish Energy Agency, 2003).

Figure 3.1 shows the development of the biofuel utilization in Sweden. Solid biomass, peat and waste supplied 98 TWh of energy in 2002, which compares with about half as much in 1980. Biomass is used in the forest industry, district heating and single-family houses for provision of heat and power through a variety of generation and end-use technologies. Despite the progress already achieved, the existing biomass resource base allows further development of bioenergy in the country. National estimates have indicated that an annual energy potential of approximately 160 TWh could be reached by 2010 using Swedish biomass sources only (Lönner et al., 1998).

[1] Considering end-use only and excluding all losses.

Taxes and investment grants have played a decisive role in enhancing the competitiveness of bioenergy in Sweden (Bohlin, 2001). Fossil fuels have been taxed in the form of CO_2 taxes, sulfur taxes, NO_X taxes and the general energy tax[2]. Investment grants have also been provided for the establishment of bioenergy plants. Since energy and environmental taxes have distinguished between types of users and energy carriers, certain segments have been particularly encouraged. This has been the case for district heating. In 2002, biomass responded to 35.5 TWh, or approximately 65 per cent of the total district heating consumed in Sweden, compared with a very marginal contribution two decades earlier.

The share of district heating in Sweden is high by any international comparison. Nevertheless, there is a significant potential to use more district heating based on biomass, if such systems can be established in areas with more sparsely distributed heat demand. One-third of the Swedish single-family houses are still being heated with electricity. Further development of the district heating system can help release electricity from the heating system for use elsewhere, while also making the country's total energy system more efficient. The conditions through which this may become possible are further discussed by Sandberg and Bernotat in Chapter 8.

The conversion of heating systems away from electricity in single-family houses has occupied the attention of energy planners in Sweden for a long time but the progress achieved has been limited. Besides convenience to end-users, electricity prices have been relatively low in the past, while initial investments to shift systems are high for a household. Getting a bioenergy equipment installed has proven time consuming for potential users who are not familiar with the new technologies, biofuel distribution chains, permissions needed and incentives available. Beyond that, users have been reluctant to install bioenergy systems due to the more intensive maintenance required in comparison with fossil- or electricity-based heating systems. Meanwhile, heat pumps have gained popularity. Today, however, bioenergy is probably the most cost-efficient alternative for heating single-family houses in Sweden. The market seems to have reached a critical mass of installed units to prove this and, with further incentives being provided, bioenergy use in single-family houses is bound to increase.

The forest industry is the major producer and user of bioenergy in Sweden. Of the 51 TWh of biomass used in the industrial sector in 2002, about 80 per cent were used in the pulp and paper industry. However, this should not be understood as a result of energy policy incentives, as there are synergies that favor bioenergy utilization in the sector. Pulp and paper production is very energy intensive, and the annual use

[2] It is not clear if the NO_X tax has favored biomass-based compared to fossil-based energy. The general energy tax, however, has had its major impact on the private heating market, since it does not apply on electricity production and the industry sector at large.

of energy in the sector can vary significantly from year to year due to variations in markets for forest products.

Though Swedish policies have been quite successful in enhancing the use of biomass for heat production, the same cannot be said about the fuel mix in power generation and transport. Only 6 TWh of electricity and 0.5 per cent of the total amount of fuel used in the transport sector came from biomass in 2002. When it comes to electricity, the same incentives provided for heating were not considered possible. It is increasingly difficult for Sweden to tax electricity generation very differently from neighboring countries due to the integration of electricity markets. Moreover, it is believed that the competitiveness of energy-intensive segments of the Swedish industry would be severely affected if they were obliged to pay electricity prices much higher than today. For that reason, these sectors have been exempted from the various taxes that might otherwise have favored a fuel shift.

As for the transport sector, the main barrier to change has been the lack of alternatives, that is, a competitive renewable alternative to fossil-fuel-based transport. A competitive alternative has to cover the whole chain from biofuel production to biofuel adapted vehicles, to infrastructure for storage and distribution. However, ethanol has already been introduced in Sweden mixed with gasoline. Two factories are producing 57 000 m^3 of ethanol for use as transport fuel in Sweden. Research on alternative motor fuels has received continuous support since 1975 (Zacchi and Vallander, 2001). A pilot plant to produce ethanol from lignocellulosic materials such as agricultural wastes, wood and municipal waste has been established recently.

3.3. IMPORTANT DRIVES AFFECTING BIOENERGY UTILIZATION

Conditions for the implementation of different energy systems vary from country to country. For example, the natural resource availability, existing infrastructure and the types of services on demand are likely to define technology choices and system options (Roos et al., 1999). Policies and priorities may also imply more or less favorable conditions for given technologies and solutions, setting the development path, boosting or canceling opportunities. A technology breakthrough, major political or social events, as well as natural disasters may also influence choices significantly, thus affecting the direction of development.

Rather than trying to cover the whole spectrum of factors that may affect the evolution of bioenergy systems, we refer to three major drives that are likely to have a direct influence on the design of strategies and policies for bioenergy utilization in Sweden. These drives are internationalization of the bioenergy sector, integration of bioenergy systems with other production systems, and the mainstreaming of bioenergy as a major energy source. We believe that these trends will also affect the development of bioenergy use in Europe at large, not least due to the present energy policies being applied in the region, and the recent accession of ten new countries into the European Union.

Table 3.1 shows how these drives relate to specific actions and opportunities when it comes to enhancing the resource base, further integrating bioenergy with forest industries, pursuing climate policies and business opportunities. We discuss each of these drives and groups of actions separately.

Internationalization of the bioenergy segment

The internationalization processes that affect the bioenergy segment more directly are reflected in (i) the development of markets for biofuels, (ii) definition of policies and actions to favor bioenergy options, and (iii) research and development.

Traditionally, biomass for energy has been harnessed and used locally, but trade in biofuels is, in fact, expanding rapidly, boosted by commercial opportunities. These opportunities are anchored in well-established systems and are on increasing demand. In Sweden, the imports of biofuels have increased from a volume corresponding to less than one TWh in the beginning of the 1990s to 6–8 TWh today. Further market development in the near future shall be strongly affected by national and regional policies.

Policy making reflects the internationalization of the bioenergy sector. Various EU directives deal directly or indirectly with energy issues, and some are particularly significant for the development of bioenergy e.g. directives dealing with energy taxation, waste incineration, combined heat and power production (CHP), and motor fuels (European Commission, 1997, 2001). Standards and specifications are also the subject of common projects within the EU, e.g. European standards for solid biofuels and for solid recovered fuels are being developed (see also Thrän et al., Chapter 11). These standards aim at improving conditions for trading biofuels. The development of a common policy framework for the EU is also in line with international agreements such as the Climate Convention and Kyoto Protocol, and may gradually lead to common strategies and legislation for bioenergy.

Internationalization is also observed in R&D. The EU aims at a better coordination of research and development as a means to make Europe more competitive, and this is being promoted through specific schemes to distribute EU research funds. It is believed that increasing complexity implies high costs for further developing bioenergy systems, and small countries such as Sweden cannot cope with the task alone. Obviously, such coordinated efforts find barriers particularly due to the large variation in the progress achieved by the various EU countries when it comes to understanding and using bioenergy systems.

Integration of bioenergy systems with other production processes

A more rapid and effective development of bioenergy, to reach more volume and importance in the European energy matrix, and offer the reliability and cost efficiency commercially required, demands integration and coordination. The bioenergy sector needs to be more integrated with other segments of the energy sector, for example, as part of strategies to secure the energy supply, or as a means to make bioenergy a more competitive alternative.

Bioenergy can be better motivated when integrated with other business sectors and industrial processes. For example, there is a significant potential for synergies through increased integration with the forest industry (see e.g. STFI, 2000; see also the discussion in Chapter 7). Bioenergy generation companies need vertical integration of the fuel chain to guarantee quality biofuels derived from waste handling and forestry activities. The sector also needs to forward integration in consumer markets in order to exploit the full potential and qualities of bioenergy. But there are barriers to such integration. For example, biofuel and bioenergy production are at the margin of core activities of most forest companies (e.g. Ling, 1999). Other non-technical barriers include issues related to the distribution of business ownership, as well as the sharing of responsibility for management and risks.

There is integration potential also with sectors such as waste management and rural development, which conventionally belong to other departments. Such integration requires a coordination of policies, planning and development, and strategies for marketing bioenergy. This implies coordination of public and private actors from different business spheres. The experience of Enköping provides an example of how this can be made possible (see Fact box 3.1). In fact, the potential to contribute to environmental benefits, new business opportunities and regional development while providing efficient energy services can be crucial in assuring continued support for bioenergy and further progress in this area.

Bioenergy as a mainstream alternative

Modern bioenergy options have not been typical choices when the supply of heat, power and liquid fuels are contemplated. However, as bioenergy evolves from being a peripheral alternative to becoming a mainstream player in the energy sector, conditions for designing strategies for the segment change significantly. The framework within which bioenergy shall compete becomes more and more apparent.

Not least, there is the difficult question of where in the supply chain the available biomass resources are optimally utilized. Key actors in this competition are positioned and organized, and this includes biomass producers, the various links of the energy industry, and the combustion sector, as well as the agencies and institutions involved in energy research.

Fact box 3.1. *Integrated system for bioenergy, water treatment and regional development – the experience of Enköping*

Enköping is a small town located in the middle of Sweden, about 70 km from Stockholm. In 1972, the municipality founded Enköpings Värmeverk to produce and distribute heat to the local community. Today, 20 000 people live in the urban area of Enköping. Thirty-six employees work with the operation and maintenance of the 76-km-long district heating network that provides the majority of residential, commercial and industrial buildings of the town with heating services. Some 220 GWh of heat are distributed to more than 1400 customers every year, of which some 1100 are single-family houses.

Originally, the three boilers at Enköpings Värmeverk operated on oil and propane gas but, nowdays, the heat production relies mostly on biofuels. One of the boilers has been converted for burning wood powder and has an output of 20 MW. It produces around 15–18 per cent of the requirement for district heating and runs from the middle of May to the middle of September. It also runs during the winter at low temperatures. In the colder part of the year, from the middle of September to the middle of May, the biofuel-fired CHP plant, Ena Kraft, meets the demand for heating. The CHP plant produces around 80–85 per cent of the yearly consumption of district heating in Enköping. The plant uses bark, sawdust, residues from logging operations and salix.

Local farmers are planting the salix that is heating homes and industries in Enköping. As part of an innovative project, the municipal council has financed the salix plantations, and a leasehold agreement has been established with each individual landowner. Residual ash from the power plant, approximately 1500 tons per year, is mixed with sludge-water and distributed to local salix plantations. The water used for irrigation passes the conventional purifying process prior to being pumped into ponds. After filtration, the water is distributed throughout the 80-hectare salix plantation. In this way, approximately 60 tons of nitrogen/hectare are dispersed every year.

This bio-cyclical solution that utilizes nutrients from sewage treatment and ashes from energy generation to grow energy crops has been the result of a cooperation coordinated by the County Council and involving local authorities, the power plant and the farmers in Enköping. Nitrogen and phosphorous effluence, that would otherwise pollute the Lake Mälaren and the Baltic Sea, is being used to fertilize energy plantations, reducing the harvest time by 25 per cent. It is understood that the conditions for emissions of heavy metals comply with the prevailing limits and regulations. Salix is being used to help purify wastewater emanating from private septic tanks and the municipal sewage plant.

Prepared with the cooperation of Eddie Johansson, Enköpings Värmeverk

Biofuels and combustion equipment of various types and sizes are becoming more and more standardized. The transactions between buyers and sellers are also to a great extent, regulated through standardized contracts. This indicates a more mature phase of the bioenergy business segment. This new phase is strongly anchored on the experience of countries like Sweden and a few others, which now pave the way to a larger dissemination of bioenergy options.

Bioenergy is becoming a mainstream alternative, and more of a standardized business chain characterized by its bulkiness and management complexity. In Sweden, and probably in most of Europe where energy markets are mature, economies of scale are likely to become more important for the commercial attractiveness of the business. In contrast, the small-scale use of biomass, particularly in Swedish rural homes, will remain very competitive due to the low cost of fuel and difficult taxation of these privately generated services.

3.4. FOUR MAJOR TASKS IN THE DEVELOPMENT OF BIOENERGY IN SWEDEN

The fact that bioenergy has many links to other sectors and activities creates both problems and opportunities for its dissemination. There are definitely opportunities for developing bioenergy applications within win–win frameworks, combining solutions for different interests and functions in society. However, the multitude of issues involved often creates complex and time-consuming political processes.

Finding and developing political, technical and economic win–win niches for bioenergy is, therefore, the key to effective implementation strategies. We need to detect major functions of bioenergy in society, and the drivers and obstacles connected to these functions, as well as interlinks between them, to be able to explore the synergies. In this section, we discuss some major tasks related to these functions, and ways through which bioenergy can be further promoted in Sweden. We continue to follow the structure presented in Table 3.1. Again, the list is not meant to be exhaustive, but rather points to some key aspects that need particular attention.

Obviously, the issues discussed here may weigh differently in different countries, and other issues may be more important than these. For example, in Sweden, integration with the forest industry is a wise focus in face of the size and importance of forestry activities. In countries where agriculture is a major economic sector, the development of a competitive nonfood activity to diversify the economy may provide good synergy effects and qualify as an essential issue. In many developing countries, rural development may receive particular attention and be a major argument in favor of bioenergy.

Enhancement of the biomass supply source

As pointed out before, bioenergy is becoming a mainstream energy alternative. But, while the business structure for bioenergy utilization is well tested and has proven efficient to this point, a number of issues need to be addressed as the expansion of bioenergy systems is contemplated. Such issues refer to the availability of biomass resources, the type of services demanded and the best form to provide them utilizing bioenergy, as well as the policies and institutional base needed to ensure that biomass will remain an attractive source of energy.

Further development of the biofuel base can be effectively achieved in the short- and medium-term based on the existing biomass potential. According to Lönner et al., (1998), the potential supply of biomass in terms of forest fuels, waste and imports within a feasible cost range are greater than the foreseeable demand within Sweden. Although this study provides an indication of the resource availability, the validity of a study focused on Swedish needs only may be questioned. Truly, Sweden has been a forerunner in the use of modern biomass technologies, but the whole European Union is contemplating bioenergy options. Thus the resource availability should be analyzed within a broader context. The trade with biofuels is increasing with major streams from the Baltic countries and Russia to Scandinavia and the Northern parts of Europe. However, with increasing demands from different countries, competition for biofuels is likely to increase.

The next leap in the use of bioenergy in Sweden will most certainly be in combined heat and power production (CHP). In this area, Sweden is actually behind Finland and Denmark. The large availability of electricity from nuclear power plants in the past decades has allowed low electricity prices on the market, providing a disincentive for new CHPs. Also the structure of the tax system and the environmental legislation in Sweden have played a role in the development observed. However, the enlargement of the power production capacity in Sweden shall count heavily on CHPs. Given the present policies and tax structure, for example to curtail greenhouse gas emissions, these CHPs are most likely to be fueled with biomass. Meanwhile, the penetration of biofuels in the transport sector shall proceed at a lower speed until perhaps a breakthrough is reached in about ten years.

When it comes to development of the biomass resource base in the next ten years, we can single out four major issues that Sweden needs to address.

- *Trade with bioenergy technology, know-how and biofuels*
  The internationalization of the bioenergy sector leads to increased trade with biofuels, while also improving markets for bioenergy equipment and know-how. The challenge is to understand the underlying conditions and incentives behind existing and potential trade patterns to be able to exploit the trade in bioenergy-related products in an effective way.
- *Transnational energy companies*
  The challenge is to understand the structure of decision-making, and the strategic considerations of emerging transnational energy companies so that they can be made proactive in promoting bioenergy options.

- *Integration of bioenergy with other socioeconomic sectors*
  Successful bioenergy projects integrated with other sectors and functions are needed in order to achieve a general support for bioenergy from society at large. The challenge lies in facilitating the creation of a wide cluster of actors that can develop and manage complex bioenergy projects.
- *Decreased tax advantage*
  As bioenergy increasingly becomes a major supply source of energy, the reasons for strong tax advantages decreases. The challenge lies in the development of bioenergy systems that are resilient to more competitive conditions.

Systems integration within the forest industry

The role of the forest industry in the development of bioenergy in Sweden can hardly be overstated (Hillring et al., 2001). Forest industries are the largest users of bioenergy today, and it is in this sector that the largest potential for bioenergy production still exists. In fact, there is significant potential for energy surplus through better process integration within these industries, a surplus that can serve to supply external users.

For a long time and with varying enthusiasm, bioenergy has been discussed as the third pillar of the forestry sector, together with sawn wood and pulp-and-paper production. In the beginning, the discussion focused particularly on the potential competition for biomass between pulp producers and bioenergy users. After many years of research, the discussion has become broader and more sophisticated today including issues such as:

- the complementarity between forest production for timber and pulp wood on the one hand, and for energy on the other hand, in terms of silviculture, logistics and overall economy of the forest production;
- the opportunities for increased energy efficiency in the pulp and paper production;
- the possibilities to combine and integrate pulp and paper production with the production of upgraded biofuels.

In fact, there are a number of cost-efficient measures to generate more biofuels and bioenergy in connection with various activities of forest industries. Recent studies show that, by using the best commercially available technology, the pulp and paper industry can make a great amount of biofuels available to the market, if only energy efficiency is given a high priority. A tentative estimate is that, within 20 years, the Swedish pulp and paper industry can produce the equivalent to 25 TWh of biofuels annually. Approximately half of that energy will be used internally to increase production capacity, while the other half can be made available to the market (STFI, 2000). Gasification of black liqueur alone has the potential to double the power generation in the sector once it reaches a stage of commercial breakthrough.

Upgraded solid biofuels such as pellets are mainly produced from the by-products of saw mills. In Sweden, only a small fraction of the net annual potential of some 35 TWh in the form of by-products is upgraded to biofuels. The largest portion of the by-products is used either to meet the internal energy demand of the industries, or as raw material in the pulp industry. In the long run, however, depending on how the price relation among various products evolves, it is realistic to expect that saw mills will use solid biofuels of lower quality to meet internal energy needs, and their own by-products to produce other fuels, e.g. pellets, thereby better exploiting the value and economy of the by-product.

In the long run, the conditions exemplified so far may lead to a situation in which a cost-efficient bioenergy production well integrated with forestry, pulp and paper and sawn wood industries will become an important competitive factor for the Swedish as well as other countries' forest industries. When it comes to exploring the benefits of such an integration of production processes in the next ten years, we single out three major issues that Sweden needs to address.

- *Develop national biomass for energy production systems*
  In the short run, the international market for biomass will provide a surplus of biomass at competitive prices. The challenge is, under the present hard market competition, to develop national systems for biomass production for energy purposes that prove competitive in the long run.
- *Combine and integrate production of pulp and paper and upgraded biofuels*
  Combined and integrated production of pulp and paper and upgraded biofuels is very promising from the energy efficiency perspective. The challenge lies in the development of production systems that create win–win synergies worth exploring, and which do not interfere too much in the main production line of e.g. pulp and paper.
- *Ash recirculation*
  As bioenergy increasingly becomes a major supply source of energy, the biomass production cannot be carried out in a way that jeopardizes the long-term production capacity of the forestland. The challenge is to facilitate the development and utilization of ash recirculation systems in terms of organization, cost-sharing and management.

Reduction of CO_2 emissions

Bioenergy can be used as a means to reduce CO_2 emissions, thus helping to hamper the increase of CO_2 concentrations in the atmosphere. This can be accomplished through the substitution of fossil fuels, or substitution of materials whose production processes generate large CO_2 emissions, e.g. concrete, steel. Moreover, carbon can be sequestered in biomass in the form of carbon sinks. Also soils have a role to play in carbon sequestration. How bioenergy can help address the global climate problem is further discussed in other chapters of this volume (see also Chapters 2 and 12). In this section, we only briefly address how the climate problem is affecting the development of bioenergy in Sweden.

In Sweden, the climate problem has been acknowledged not only as a great international challenge but also as a national driver in the development of bioenergy strategies. Though Sweden has negotiated an increase of national greenhouse gas emissions of 4 per cent within the European Union, the short-term goal is to reduce emissions by 4 per cent in the first commitment period of the Kyoto Protocol. The goal shall be met without using sinks or the flexible mechanisms.

The strategic issues connected to climate change include trade-offs between long- and short-term objectives. Short-term cost-effective measures may lead to severe impacts and, consequently, require expensive measures in the long run. Thus, should we change the energy system today or tomorrow, or more precisely, which part of the energy system can be changed on a cost-efficient basis today and tomorrow, respectively?

When it comes to measures to reduce CO_2 emissions in the next ten years, we can single out two major issues that Sweden is addressing that affect bioenergy more directly.

- *International trade with climate-related products*
  International trade with climate-related products is being established in the form of emissions trading, joint implementation and clean development mechanism. The challenge is to find ways to exploit the so-called *climate bonus* in order to promote bioenergy know-how, generation and use.
- *Complex systems, institutions and structures*
  Integrated bioenergy systems are complex and are affected by a number of drivers simultaneously. The institutions and structures of the international climate regime are under construction. The challenge lies in the development of an incentive structure that promotes climate-related products, while also contributing to dismantle barriers to bioenergy – both nationally and internationally.

Development of a competitive bioenergy industry

In general, public spending on R&D aims at creating or maintaining favorable conditions for the development of a competitive industry. Investments in bioenergy know-how are no different. A major argument for being a forerunner in the development and use of bioenergy technologies is that this will foster the accumulation of know-how, and support a competitive industry. Huge markets are being envisaged internationally. In addition to Europe and North America, China, India and other parts of Asia and South America are often discussed as major markets for bioenergy technology and know-how.

Thus trade related to the bioenergy sector includes equipment, know-how and fuels. Equipment available for export from Sweden today includes pellet burners and forest fuel harvesting equipment. The know-how can be provided through consultancy related to biomass gasification and forest fuel logistics. The exports related to equipment and know-how are more at hand than different types of biofuels. However, some upgraded biofuels may be exported from Sweden in the long term. In general, development towards more diversified trade is observed, with countries importing some biofuels and exporting others depending on where the country has its competitive edge.

Parallel to the trade evolving between industrialized countries, an interesting window of opportunity has opened for export to emerging developing countries. Favorable conditions in that context are observed in countries like China, India, Brazil and Chile. Environmental issues, particularly related to CO_2 emissions, are also contributing to foster these new markets for bioenergy technologies and services. At the same time, tropical countries with large availability of land can become important producers of biofuels for an international market.

In addition to the goal to develop a competitive bioenergy industry, there is the need to promote regional development and guarantee the security of energy supply. Bioenergy investments are suitable to promote regional development, not least because the biomass production is per definition geographically spread out. The rationale of the regional development dimension can be explained in two basic ways at the macrolevel.

- Bioenergy is to a large extent used in urban areas but produced in nonurban areas around the country. The increased use of bioenergy will allocate resources from urban areas for investment in rural areas where jobs will be created. These resources would otherwise be transferred to other regions abroad i.e. to pay for fossil fuels.
- Increased use of bioenergy will increase the value of the biomass that is produced throughout the country, mainly in the form of forest residues and by-products from the forest industry sector. These resources will generate value to various regions of the country, allowing distribution of the gains.

When it comes to measures to foster the development of a competitive bioenergy industry, we can single out three major strategies for Sweden.

- *Project clusters*
  On an international market for bioenergy systems including biofuels, equipment and know-how, it is crucial to be able to rely on a critical mass of resources and

knowledge. The challenge is to facilitate the creation of competitive bioenergy industrial clusters with significant Swedish participation.

- *System integration*
  There is a great potential for further development of bioenergy through better integration with the forest industry, waste management etc. The challenge lies in the development of the integrated systems as in the example of Enköping (Fact box 3.1).
- *Development of businesses along with bioenergy know-how development*
  Public-funded bioenergy projects have not delivered as many new products, and fostered as many new companies as was hoped for. The challenge lies in trying to develop a business environment capable of defining bioenergy products and services that can generate more economic returns.

3.5. CONCLUDING REMARKS

There is a large need to expand the energy-supply infrastructure base in the world in the coming decades. IEA (2003) estimates that the global energy system requires investments of the order of 16 trillion USD between 2001 and 2030. This is obviously a tremendous challenge involving engineering, financial and environmental dimensions. But it is also a life time opportunity for the bioenergy sector to provide alternatives that are competitive and beneficial in many respects.

The exploitation of bioenergy opportunites involves a complex and multi-functional challenge. We need to pick the bioenergy segments and applications that can serve to develop win–win solutions in cooperation with other sectors of the economy. In doing this, bioenergy will benefit from other sources of funds, not to finance the energy infrastructure per se, but to make bioenergy a more cost-efficient and sustainable solution, with multiple benefits for society. This will help create market dynamics around bioenergy as we have never experienced before.

Various factors contribute to make bioenergy applications a significant part of the coming energy-supply investments around the world. In Sweden, as well as in the whole of Europe, bioenergy shall take a final leap towards becoming a substantial and reliable supply source of energy within the coming ten years. Although this will demand huge amounts of biomass for energy, it will also further the development of technology and know-how to create a robust infrastructure system for the whole energy supply and use chain. In addition, the policy framework will be decisive in strengthening the political intentions and creating the necessary business environment to allow that to happen.

With opportunities, the pressure to deliver also comes. In the coming ten years, bioenergy experts and entrepreneurs will have to exploit and realize more of the commercial potential of bioenergy products and services in order to match the competition of other renewable energy options. In other words, it is time for the bioenergy sector to deliver at a larger scale and on its own merits!

REFERENCES

Bohlin, F. (2001) The making of a market – supply- and demand side perspectives on institutional innovation in Sweden's wood fuel use, in *Acta Universitatis Agriculturae Sueciae*, Silvestria 232, Uppsala, Sweden.

European Commission (1997) Energy for the Future: Renewable Sources of Energy – White Paper for a Community Strategy and Action Plan, COM(97)599 final, European Commission, Brussels.

European Commission (2001) European Climate Change Programme – Long Report, European Commission, Brussels.

IEA (2002) World Energy Outlook: Assessing Today's Supplies to Fuel Tomorrow's Growth, Paris.

IEA (2003) *World Energy Investment Outlook 2003*, Paris.

Hillring, B., Ling, E. & Blad, B. (2001) The potential and utilization of biomass, in *Building Sustainable Energy Systems – Swedish Experiences*, Ed. Silveira, S., Swedish Energy Agency.

Ling, E. (1999) Skogsbränslet och Skogsföretaget – en institutionell integration (Forest Fuels and Forest Industries – an Institutional Integration), Report No 3, Department of Forest Management and Products, SLU, Uppsala, Sweden.

Lönner, G., Danielsson, B-O., Vikinge, B., Parikka, M., Hektor, B. & Nilsson, P-O. (1998) Kostnader och tillgänglighet för trädbränslen på medellång sikt (Costs and Availability of Woodfuels in the Medium Term), SIMS, Rapport nr. 51, SLU, Uppsala, Sweden.

Roos, A., Graham, R., Hektor, B. & Rakos, C. (1999) Critical factors to bioenergy implementation in *Biomass & Bioenergy*, **Vol. 17**(2), Elsevier, pp 113–126.

Silveira, S. (2001) Tranformation of the energy sector, in *Building Sustainable Energy Systems – Swedish Experiences*, Ed. Silveira, S., Swedish Energy Agency.

STFI (2000) Ecocyclic Pulp Mill, Final report KAM1, 1996–1999, Stockholm.

Swedish Energy Agency (2002) Växande energy – bioenergin i Sverige, en marknad i utveckling (Growing Energy – Bioenergy in Sweden, a Market under Development), Eskilstuna.

Swedish Energy Agency (2003) Energy in Sweden: Facts and Figures, Eskilstuna.

Zacchi, G. & Vallander, L. (2001) Renewable liquid motor fuels, in *Building Sustainable Energy Systems – Swedish Experiences*, Ed. Silveira, S., Swedish Energy Agency.

Chapter 4

Dissemination of Biomass District Heating Systems in Austria: Lessons Learned

Christian Rakos[1]

4.1. DISTRICT HEATING IN AUSTRIA

Energy from biomass provides about 13 per cent (130 PJ) of all Austrian primary energy consumption today. The greatest part of this bioenergy use (60 per cent) can be attributed to traditional stoves and boilers fired with wood logs. Small district heating plants have, however, gained increasing importance in the last 20 years as providers of domestic heating in rural areas. These plants use wood chips, industrial wood wastes and straw as fuel, and provide about 5 PJ of energy per year. By 2001, more than 600 Biomass District Heating Plants (BMDH) had been established in the country.

The advantages of biomass district heating as compared to traditional heating systems were well known in Austria. BMDH would eliminate fuel handling at the individual level, allow the provision of continuous heat, and reduce emissions significantly as the individual heating systems were predominantly old and technically poor. Despite these advantages, the introduction of BMDH was by no means an easy process. It was only successful due to a unique combination of top down policies and local bottom up initiatives. Top down policies included financial incentives and the establishment of organizations focused on the management of the introduction process.

The Austrian experience in introducing district heating constitutes a relevant case study on technology dissemination. It allows for a close observation of the interaction between driving forces for innovation and barriers that need to be overcome in the process. It turns out that a combination of technological performance and socio-economic factors has been the key to the successful dissemination of district heating

[1] This chapter includes results of the doctoral thesis of the author: "Fünfzehn Jahre Biomasse-Nahwärmenetze in Österreich" (Fifteen years of biomass district heating in Austria). Technische Universität Wien, 1997. The thesis was written within the EU funded project Pathways from small scale experiments to sustainable regional development, EXPRESS PATH, CEC Contract No EV5V-CT92-0086.

Bioenergy – Realizing the Potential

in Austria. As part of a systemic management, technology introduction was accomplished paying particular attention to the social system in which it was embedded. Supportive policies have played a critical role and included a multitude of measures. Particularly in the early phase of technology diffusion, dedicated institutions managed the day-by-day initial difficulties and accelerated the learning processes through continuous communication and feedback.

4.2. THE DIFFUSION OF BMDH IN AUSTRIAN VILLAGES

The scheme of a BMDH system is simple. A big furnace fueled with biomass heats water that passes through a pipe grid and supplies energy to heat individual houses in a village. Austrian villages with BMDH plants usually have between 500 and 4000 inhabitants and are of predominantly rural character. Accordingly, the size of BMDH plants varies between a few hundred kW and 8 MW, with corresponding grids between 100 m and 20 km. About two-thirds of all plants have a power of less than 1500 kW. Plants larger than 800 kW typically supply whole villages, while smaller plants may heat only a few larger buildings in the village center. Most plants were built in Lower Austria, Upper Austria, Salzburg and Styria. Presently, there is an obvious saturation of the market of village heating systems and a sharp increase in what is called microgrids.

During the first phase of the technology introduction and until 1984, private companies were the predominant developers and operators of BMDH plants, mostly sawmills. They were followed by municipalities and farmers' cooperatives, which are now operating a great majority of plants. Utilities became more interested in BMDH in the 1990s but were rather cautious in setting up projects. In some cases, interesting forms of joint ventures with farmers' cooperatives were established. The utilities took the role of developers taking advantage of their professional, technical and management know-how, while farmers took the role of operating the plant and arranging for fuel supply. It is important to mention that the general dominance of farmers as developers and operators is related to the enhanced availability of subsidies for this group.

4.3. TECHNOLOGY PERFORMANCE AND QUALIFICATION OF PROFESSIONALS

Four distinct aspects of technology performance influenced the technology diffusion quite differently: performance of the central heating plant with respect to the reliability of the operations, performance with respect to emissions, the technical interface with heat consumers, and overall systems efficiency.

Reliable operation was a major preoccupation to operators in the beginning. It did not influence diffusion significantly because the proud plant owners were secretive about operation problems. Due to their technical versatility, they managed to overcome daily problems. Customers usually do not notice central plant failures of a few hours only, due to the heat capacity of the hot water in the grid.

A close follow-up of the problems faced in the plants was carried out with the equipment providers with the help of the supporting organizations and the so-called *technology introduction managers*. This contributed to learn-by-doing and led to the necessary technological improvements for reliable plant operation. Figure 4.1 shows the effect of the process of technological learning. The percentage of plants with serious operation problems dropped sharply after the initial years of the technology dissemination. By the mid-1980s, almost one-third of the plants were problem-free.

The second aspect of technology performance was related to emissions from the heating plant. In the early 1980s, emissions played a secondary role in the public perception of technology performance. Nevertheless, a publicly funded R&D program was carried out and this led to the deployment of emissions mitigating equipment employing continuous power control, electronic combustion control and, in the early 1990s, flue gas condensation. The attention given to the issue of emissions was worthwhile as the technological improvements achieved favored further dissemination of district heating plants, particularly in the 1990s when environmental awareness became more widespread.

The greatest threat to the district heating technology diffusion came from deficiencies in the technical interface between the district heating grid and the

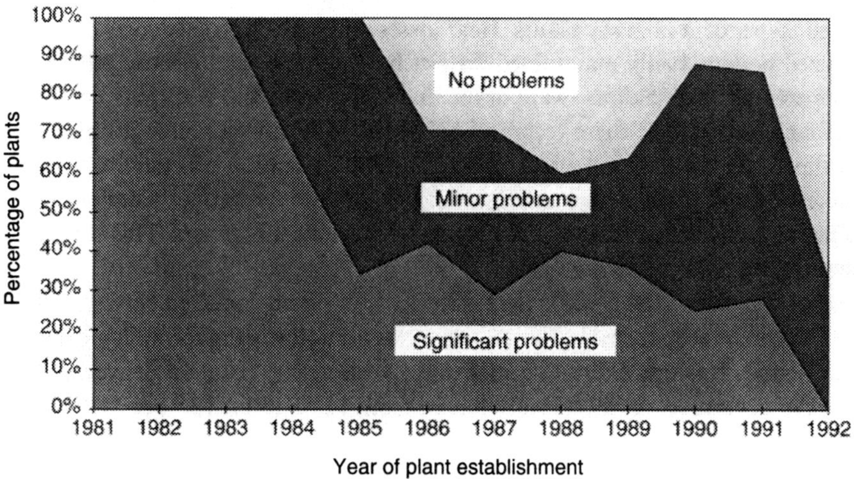

Figure 4.1. Operational problems in BMDH plants 1981–1992.

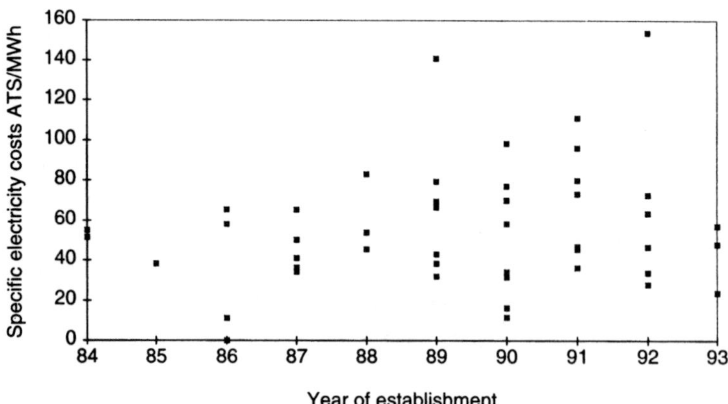

Figure 4.2. Electricity costs in plants installed between 1984 and 1993.

individual house-heating systems. Failures in this subsystem directly affect consumer comfort. This interface, including for example the heat exchanger and pipes, is not a part of the BMDH plant. Normally, the equipment was planned, installed and maintained by independent local plumbers, who did not have any particular experience with district heating. In some cases, plumbers repeated the same mistakes in many installations in a single district, which seriously affected the local reputation of biomass district heating. It so happened that plants could not be established in neighboring villages due to the bad reputation created.

The last technical aspect is the overall plant performance. While considerable R&D efforts were made to minimize emissions, overall technical performance of the system was largely neglected. Even today, the annual system efficiency is only around 50 per cent in many plants. Heat losses and high electricity costs occur due to oversized boilers, badly maintained district heating pipes and electric pumps, vents, etc. Operators were seldom aware of the high heat losses and electricity costs of their plants and understood these technical problems to be an economic problem related to competition with low oil prices. Since no public attention was paid to monitoring technical overall performance and as the operators did not properly understand such problems, no feedback reached the responsible technical planners. This slowed down technological learning considerably.

Figure 4.2 shows the specific electricity costs of plants installed between 1984 and 1993. As can be seen from the figure, there was no noteworthy technological learning in this period. It is significant to notice that the electricity costs for delivering 1 MWh of heat to customers varies widely, that is, from 15 to 150 ATS[2] per MWh between the best and the worst plants.

[2] 1 € = 13.76 ATS.

It is difficult to quantify what impact the deficiencies in planning and operation had on the dynamics of technology diffusion. They certainly led to unnecessary costs and management problems. Politicians managed to prevent open financial disasters by asking public utilities to take over plants with serious problems. Proactive policies to upgrade plants technically were only put in place 20 years after diffusion kick-off with the introduction of technical quality criteria as precondition for subsidies.

Thus lack of qualification among relevant professionals such as plumbers, planners and plant operators was a major technical obstacle for biomass plant diffusion in Austria. This is important to emphasize as there is a tendency to focus attention on the technical device per se and less so on what is considered more peripheral such as technology interface and professional skills. Feedback at all points along the energy generation and distribution chain is fundamental for technological learning. Putting in place appropriate feedback mechanisms should therefore be regarded as a central task for renewable energy management.

4.4. THE SOCIOECONOMIC CONDITIONS OF VILLAGES

A key factor in understanding BMDH development in Austria is the difficult economic situation of farmers, particularly in areas with low tourism and a declining industrial base. The majority of Austrian farmers are also forest owners. Farmers own about half of the Austrian forests and their properties are usually smaller than 40 ha. Thus farmers in areas with low rates of development are eagerly seeking alternative income sources within agriculture and forestry.

During the last 20 years, there has been an oversupply of wood in the market. Meanwhile, wood prices have decreased significantly due to competition with cheap wood imports, periodic crises in the pulp and paper industry and increased use of recycling paper. The substantial decline in wood prices was a major driving force behind BMDH development. This technology offered a chance to add value to wood that was neither suitable for sawmills nor paper production.

Thus, not surprisingly, most BMDH plants in Austria were established in peripheral regions and were motivated by the opportunity to improve local socioeconomic conditions. Normally, in places where there are economic opportunities in other sectors such as tourism or industry, the interest in establishing BMDH is low. This is true for most of the western states of Austria.

Still, it is possible to motivate the technology in other socioeconomic contexts for other reasons. There are cases of prosperous tourist villages where BMDH plants were established for reasons such as comfort, local air pollution and prestige. In these cases, more advanced technologies such as flue gas condensation were necessary to prevent any visible emissions which might bother the tourists.

4.5. ECONOMIC ASPECTS OF PLANTS

In practice, the specific investment costs of biomass district heating plants show widely differing values, which depend on local preconditions, planning competence and philosophy, operators etc. The range of costs is from 360 €/kW of installed power to 1800 €/kW with average values around 850 €/kW. The costs of the boiler amount to one-third of the total investment, and the grid another third. Figure 4.3 shows the average contribution of different plant components to the total investment costs of 80 plants investigated.

Figure 4.4 shows the composition of the operational costs of BMDH plants. While data on investment costs were available from most of the 80 plants investigated, many operators did not provide data on operational costs. Thus Figure 4.5 is based on the data of 3 plants only. Nevertheless, it does illustrate well what characterizes all plants, that is, the main cost factors in the operational costs of biomass district heating plants are capital costs and fuel costs.

To analyze the role of different factors in the economic performance of a plant, a sensitivity analysis was conducted for the specific case of a 1000 kW plant built in 1995 in the province of Styria. This plant had moderate investment costs of 640 000 €, a short district heating grid of 600 m and a heat production of 1300 MWh per year. The biomass used is a mixture of 25 per cent dry woodchips supplied by farmers (heating value 3400 kWh/ton, price 82 €/t) and 75 per cent industrial

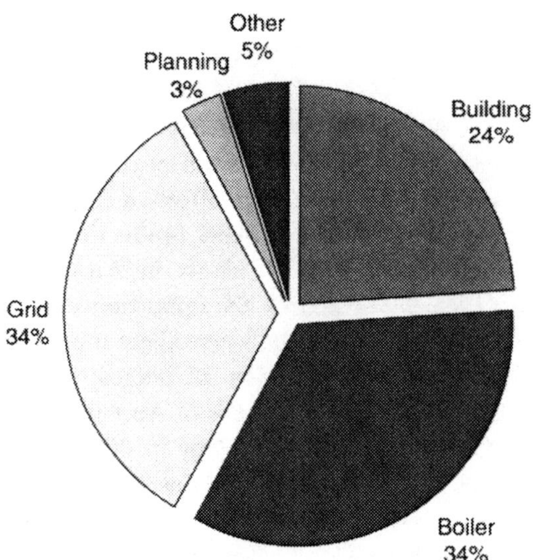

Figure 4.3. Investment costs of BMDHs in Austria.

wood chips (2800 kWh/ton, 32 €/t). The average heat price for customers was 0.063 €/kWh (all prices excluding value added tax). The capital for investment is composed of 15 per cent private capital from the members of the cooperative, 15 per cent connection fees, 35 per cent subsidies and 35 per cent agro-investment loan with 4.5 per cent interest and 15 years payback time. The calculated interest rate for privately invested capital was 4 per cent.

The dynamic model *Biowirt* commonly used for calculating the economics of BMDH plants was used to analyze how the variation of different parameters affected the amortization time of the project. It turned out that the two most critical factors for the economy of the BMDH are the heat price and the heat sales that can be achieved. A crucial precondition to make the project viable is the readiness of consumers to connect to the district-heating grid and pay a somewhat higher price than for individual heating. This readiness is in fact achievable, as biomass district heating offers significantly enhanced comfort compared to individual heating systems that are frequently in poor conditions in rural areas. Surprisingly, some consulting firms advised operators to sell heat as cheap as possible to increase sales fast – a disastrous proposal when considered in light of the findings of the sensitivity analysis.

Besides enhanced comfort, environmental protection and local self-sufficiency also play a significant role in the motivation of district-heating customers. Economic considerations play neither a central role nor are they consistent. This conclusion is

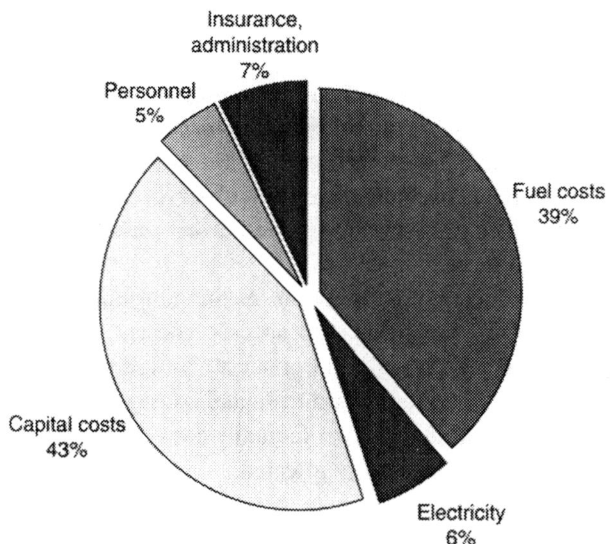

Figure 4.4. Operation costs of BMDHs in Austria.

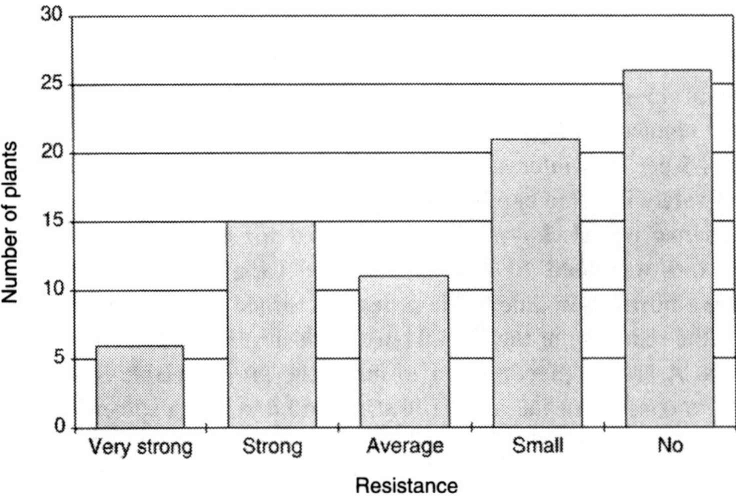

Figure 4.5. Resistance against BMDH projects in Austrian villages.

emphasized by a survey on the opinions and experiences of customers in different villages. The survey showed no consistent relationship between the economic evaluation of BMDH by consumers and the actual heat prices paid (which differed by more than 20 per cent).

4.6. THE SOCIOCULTURAL CONTEXT

A major barrier found particularly early in the innovation process was distrust of the new technology. Will it work? What will be its impact on village life? Who is going to profit from the project? These were some of the questions usually discussed for months at the village inns. Conflicts have been observed in the majority of villages where plants were installed and these were often rather serious. Figure 4.5 illustrates the level of resistance observed.

Such mistrust of new technology is by no means unusual and is often observed regardless of the type of innovation and specific context. It has to do with the cultural integrity of a society. Thus it is not simply an individual phenomenon, but also a social one. Rational economic and technical considerations will only serve to create trust if they both symbolically and factually converge with the social meaning accepted by the majority of the society affected.

Since the 1950s, rural communities in Austria have experienced the profound impacts of technical innovations in agriculture that not only completely changed the way agriculture was conducted but also changed the rural culture. New forms of life and increasing economic pressure on farmers led to social disintegration and a feeling of meaninglessness in many places. The result is suspicion regarding any form of innovation that can possibly change further or destroy local cultural habits. Parallel to that, there is a genuine desire to support initiatives that may bring new hope for rural development. The tension between these two dispositions explains the wide spectrum of reactions towards BMDH projects. The full collective support as well as vivid conflicts may be associated with such a project (see Figure 4.5). In most cases, the conflicts could be settled. However, BMDH consultants report of villages where local conflicts caused the cancellation of projects.

Two basic categories of conflicts could be distinguished in the Austrian case. The first one is related with the so-called *syndrome of acquired depression* that seems to be related to the general cultural and social disintegration of rural areas as mentioned earlier[3]. The retraction of the village economy, often combined with a long-lasting autocratic local political elite, may lead to a total apathy of the population. People have lost all hope for a better future. An innovative project not only challenges this depressed attitude but perhaps also the ruling elite. The typical attitude in such a village is distrust and rivalry. Under these conditions, the main arguments against a BMDH project are irrational or pseudoeconomic.

The second type of conflict is related to the *NIMBY Syndrome* (i.e. Not-in-my-backyard Syndrome). BMDH is nice for everybody except those who live close to the chimney and fear to be bothered by smoke or noise. This type of conflict appears quite often in places with many new settlers, usually upper class residents from urban areas seeking unspoiled nature in the countryside. These settlers are usually well organized and try to use rational or even scientific arguments to stop the project.

It is of central importance for a BMDH project that conflicts are properly and timely addressed to avoid unnecessary costs. We found that the average investment costs for plants meeting strong or very strong resistance were 30 per cent higher than for plants with no resistance. Cost increases were caused, for instance, by the necessity to change the location of the project or due to extra requirements for licensing. Lower heat sales due to the unwillingness of opponents to connect to the grid may also have a serious economic impact on the project.

The institutions that were managing technology deployment were so geared to deal with economic and technical questions that they did not address the social aspect of technology introduction adequately. Any systemic management approach needs to take this point as a key issue to avoid economic inefficiencies and limited diffusion of the technology. It is quite possible that many villages, potentially suited to receive the technology, were lost due to mismanagement of local conflicts.

[3] This issue was investigated in the course of the *Express Path* project by Kunze, G. in Cultural change and the diffusion of sustainable energy technologies. Unpublished.

This serves to explain the early decline in the rate of establishment of new BMDH systems.

4.7. THE ROLE OF POLICIES IN SUPPORTING TECHNOLOGY INTRODUCTION

The Austrian political system is such that it allows a study of the effects of different policies on BMDH deployment individually. Austria is a federal republic with nine different states. The energy policy of these states is quite different and had a profound impact on the rate of diffusion of BMDH technology. Remarkably, the majority of the BMDH plants is situated in only four states (Lower Austria, Upper Austria, Salzburg and Styria).

The comparison of state policies and their impacts shows the role of various economic incentives in the diffusion process. During the early development phase of BMDH, implementation management was of central importance. Successful introduction occurred only in provinces that established a dedicated institution or focal point that managed daily problems effectively. These institutions facilitated cooperation among all relevant actors, conducted public information activities and provided advice to local developers.

After 5–10 years of dedicated introduction management, the establishment of a plant became more of a routine process. Economic incentives established in neighboring provinces were able to learn from previous lessons and foster diffusion faster. Nevertheless, efforts to keep the development on track remain important even after twenty years since the first plant was established. The actual tasks that need to be addressed include the establishment of a program to upgrade old plants, benchmarking of plant performance, and educational activities for operators.

4.8. CONCLUSIONS

Given the complex approach necessary to get technology deployment started, two common myths regarding renewable energy can be discarded. The first says that renewable energy is primarily a question of research. The second myth says it is nothing but a question of economic incentives. Neither myth is true. Admittedly, both research and economic incentives are important ingredients. However, they must be integrated systematically taking into account the complexity of issues involved in setting up a new energy system. In other words, a systematic approach is required if the efforts are to succeed. These conclusions are also confirmed in a recent investigation where about 30 different cases of successful market deployment of energy technologies were compared (Kliman, 2001). In the majority of the cases, success was closely linked to dedicated institutions managing the innovation.

The case of biomass district heating in Austria shows the complexity of establishing a renewable energy system. It is of fundamental importance for successful renewable energy policies to avoid a simplistic economic and technical focus, and address this complexity. Resources need to be made available for a *systemic management* during the introduction of renewable energy technologies. Money invested in proper advice, monitoring of technical development, benchmarking, quality control, educational measures, and promotion based on a profound understanding of the social processes in communities is an indispensable prerequisite for success.

REFERENCES

Kliman, M. (2001) Developing markets for new energy technologies: a review of the case studies from the market barrier perspective, paper presented to IEA workshop Technologies require markets: best practices and lessons learned in energy technology deployment policies, Paris, 28–29 November.

Rakos, C. (1997) Fünfzehn Jahre Biomasse-Nahwärmenetze in Österreich (Fifteen Years of Biomass District Heating in Austria), Technische Universität Wien.

Part II

Managing Resources and Enhancing Biomass Production

Chapter 5
Managing Fuelwood Supply in Himalayan Mountain Forests

Kamal Rijal

5.1. THE IMPORTANCE OF THE FOREST SECTOR IN MOUNTAIN AREAS

In recent years, planning in the forestry sector has evolved to focus on how forest products (primarily, fuelwood and timber) can be utilized on a sustainable basis to contribute to sustainable development (FAO, 1993; Shen and Contreras-Hermosilla 1995). This is even more important in mountain areas, which are environmentally more vulnerable than the plains due to their fragile ecosystems. In addition, contradictions may exist between long-term development goals and the short-term necessities of the mountain population. Resolving this contradiction is a prerequisite for establishing the long-term vision needed to achieve changes in mountain energy systems along a sustainable path.

Sustainable development in the mountain areas depends on the capacity to develop woodfuel-based energy aiming at fulfilling the energy needs of mountain communities and increasing productivity, without jeopardizing livelihoods or depleting the forest resource base. An appropriate approach to accomplish this is to value the environment and treat it as a central feature in wood-based energy planning in the mountain areas. The choice between various forms of energy needs to be assessed, and the value of environmental stocks and flows must be accounted for, along with the role of forests as carbon sinks (Durning, 1993). The quality of life and environmental balance should be considered as important as economic growth.

In terms of fuelwood, the main concern is with resources that can be exploited for short-term benefits, which ultimately may destroy the resource base. To prioritize long-term benefits is difficult as most mountain communities lack other fuel choices. Commercial fuels and renewable energy technologies are usually not available in these areas and are not within easy reach. Even if availability of other options increases, there is still the issue of low affordability among the communities. It is therefore important to devise a mechanism of control over resources and decisions on development paths which is kept in the hands of the mountain communities themselves. The communities need to be given the power to influence the decisions

Bioenergy – Realizing the Potential

that affect their lives, which can be better achieved if they have greater control over the physical, financial and environmental capital on which they depend.

This chapter proposes to identify technological, policy and institutional options that may be feasible for the sustainable supply of fuelwood in the mountain communities of Hindu Kush, the Himalayan region of Asia[1], here called HKH region. We start by examining the management and planning efforts within the forestry sector in the region, how energy services are being met through the use of fuelwood and what the long-term implications of present practices are. We also look at lessons learnt from the implementation of forestry programs in mountain areas and propose a framework for the sustainable management of fuelwood to the benefit of mountain communities.

5.2. ENERGY SERVICES IN THE HINDU KUSH HIMALAYAN REGION

In general, per capita final energy consumption is lower in mountain areas when compared with country averages. On the other hand, the percentage of per capita energy consumption coming from fuelwood is substantially higher in the HKH region than in the respective countries. In India, for instance, fuelwood accounts for 66 per cent of the energy use in the HKH region compared to 47 per cent in the country as a whole (Rijal, 1999). This reflects the fact that mountain regions are marginalized in terms of access to commercial fuels, which makes them heavily dependent on fuelwood. This situation is worsened by the low level of efficiency in fuel utilization, which may also lead to health hazards, particularly affecting women who are the managers, producers and users of energy at the household level.

Cooking and heating are the main household energy uses in the HKH region, and a variety of traditional cooking and heating stoves fired with fuelwood is used among households. In mountain areas, demands are greater for space heating than cooking, if a comparison is made in terms of useful energy requirements. A typical example is that of Nepal where, 32 per cent of the useful energy required by the household sector in the mountains is used for cooking and 56 per cent for heating, compared with 40 per cent for cooking and 36 per cent for heating in the hill areas (Rijal, 1999). Lighting energy needs are met by kerosene and electricity, but electricity is not available in many parts of the mountains.

The energy needs of cottage industries (such as agro-processing, charcoal production, potteries, bakeries, blacksmiths, sawmills, carpenters' shop and village

[1] We refer to the 3500 km mountain range that stretches from Afghanistan in the west through Pakistan, India, China, Nepal, Bhutan, and Bangladesh to Myanmar in the east. This region is home to more than 140 million people.

workshop) include requirements for lighting, process heat and motive power. In general, the process heat requirements in facilities such as forges, potteries, and bakeries are met with fuelwood, although coal is also used extensively in the HKH region of China. Motive power requirements are met by electricity, diesel and kerosene where available, or else by human or animal labor using mechanical equipment. The use of fuelwood is widespread in agro-based facilities such as those for crop drying. The bulk of energy inputs for land preparation, cultivation, postharvest processing, and agriculture-related transport are in the form of human and animal labor. The degree of mechanization and use of commercial fuels in the mountain areas is generally low.

The pattern of energy use in the HKH region is characterized by the following (Rijal, 1999):

- Biomass dominates as a fuel, with fuelwood being the main source of energy;
- The household sector is a major consumer of energy (see Table 5.1);
- Energy demand is increasing as the result of agricultural diversification and intensification, rural industrialization, and increasing tourism;
- Energy use in mountain households varies with the household size, altitude, ethnic group, income and expenditure, land holding, livestock holding, and number of cooking stoves employed;
- The requirement for heat energy, primarily for cooking and heating, is higher than that for energy for shaft power as input to production processes;
- The demand for fuelwood exceeds the sustainable supply, and thus the process of destruction is a common phenomenon in large parts of the region;
- The cost of energy extraction is increasing;
- The availability of fuelwood is decreasing and the time taken for its collection is increasing;

Table 5.1. Final energy consumption per capita in the HKH region, by sector in selected countries 1994–1995

	China		India		Nepal		Pakistan	
Sector	MJ/Cap	%	MJ/Cap	%	MJ/Cap	%	MJ/Cap	%
Domestic	26 857	62	11 045	76	11 515	91	8163	70
Commercial	4440	10	568	4	172	1	258	2
Industrial	10 515	24	1705	12	613	5	1580	14
Agriculture	187	<1	220	1	100	<1	229	2
Transport	1216	3	1070	7	327	3	1349	12
Total	43 214	100	14 607	100	12 727	100	11 577	100

- Continuous unsustainable use of fuelwood from the forest forces rural people to use alternative biomass fuels, degrading the environment even further; and
- Availability and access to energy technologies are improving, but not enough yet to show a reduction in human drudgery (particularly of women).

Various studies have shown that there is a tendency for fuelwood energy use to decline as GNP increases (FAO-RWEDP, 1997; Ramana, 1998; Rijal, 1999). Figure 5.1 illustrates fuel preferences in relation to income, as well as the effects of fuelwood scarcity. Truly, mountain households with increased income tend to switch from fuelwood to other fuels such as kerosene, electricity or gas. However, if the more convenient alternatives are not available or if the supply is not reliable (which is common in the mountain areas), they may refrain from the upgrade. Likewise, where fuelwood is scarce, people may downgrade to lower quality fuels.

There are no readily available substitutes for fuelwood in rural mountains, but there is a clear potential for promoting energy-saving devices in these areas. However, low affordability among local populations definitely limits the dissemination of these technological options. As a result, low useful energy utilization is the

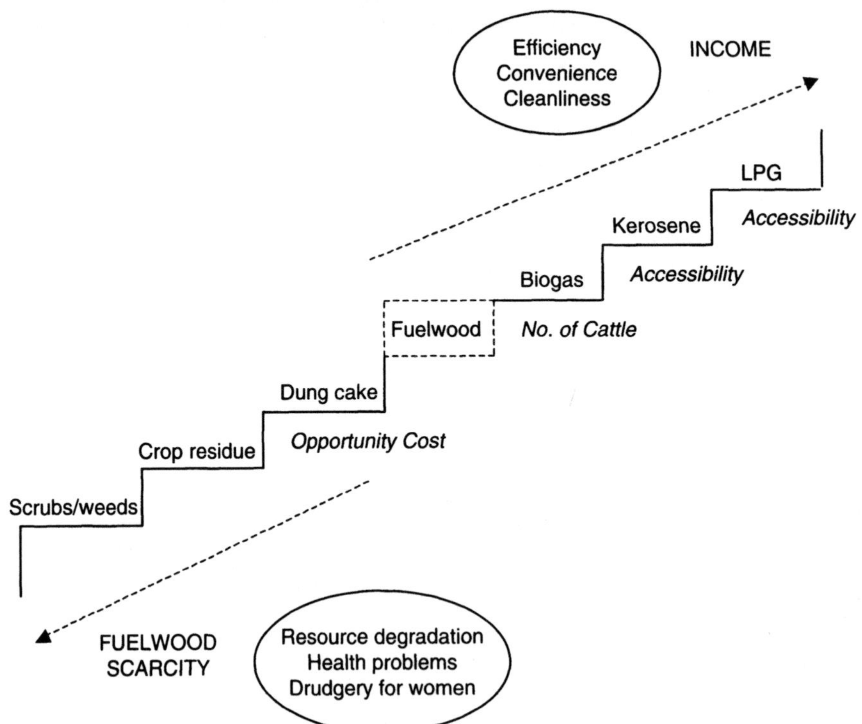

Figure 5.1. Fuel preference in cooking.

Table 5.2. Fuelwood-based traditional and new energy technologies employed in mountain areas

Demand sectors	Traditional energy devices	New options available
Cooking	Traditional stoves (3–10%) charcoal kiln (3–10%)	Mud-built improved cooking-stoves (15–20%) briquetting technology and stoves (50%) efficient charcoal kiln (25–30%)
Heating	Tripod stand (3–5%) charcoal kiln (3–10%)	Metal stoves of different designs (25%) briquetting technology and stoves (50%) efficient charcoal kiln (25–30%)
Lighting	Wooden stick of chir pine (n.a.)	
Process heat	Traditional Fuelwood Kiln (10–15%)	Efficient fuelwood kiln (25–30%) briquetting technology and end-use device (50%) efficient charcoal kiln (25–30%) Biomass Gasifiers (40%)
Motive power		Biomass gasifiers (40%)

Source: ICIMOD and CRT, 1997; and Rijal, 2001.
Note: Bracketed figures are efficiency of conversion.

rule in mountain areas, with usually less than 20 per cent energy efficiency (see also Table 5.2). Fuelwood is currently collected in the slack season at no cost other than the time and labor involved. Given widespread unemployment, the opportunity cost for the time of unskilled labor is lower than the price of fuelwood. This means that wood is likely to remain the dominant fuel in the mountains in the foreseeable future.

Traditional cooking and heating devices are prevalent in most of the mountain areas, but a variety of modern cooking and heating stoves, biomass briquettes, and gasifiers fired with fuelwood are being promoted in some selected mountain areas to aim at different end uses such as motive power, cooking, heating and lighting (see also Table 5.2). These actions still need to be supplemented with the promotion of private sector participation in technology development, institution building, and research and development.

5.3. FUEL FROM MOUNTAIN FORESTS

The choice of particular energy forms in the mountains is the result of fuel availability and access to particular energy resources and technologies at affordable prices. In this context, forest is and will remain the mainstay of energy sources in the mountain areas in the foreseeable future. Figure 5.2 shows the forest and energy linkages in the region. The contribution of fuelwood amounts to more than 80 per cent of the total energy requirement in Nepal and Bhutan, 66 per cent in India, 52 per cent in Pakistan and 29 per cent in China within the HKH region.

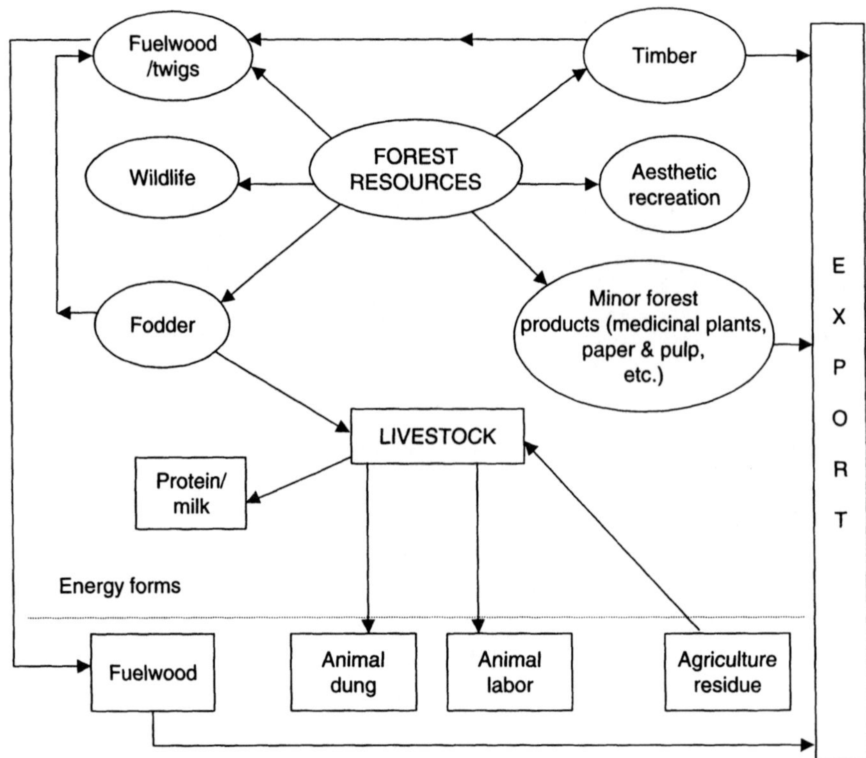

Figure 5.2. Forest and energy linkages.

There is evidence of increasing use of biomass resources other than fuelwood such as agriculture residues and animal dung as a source of energy in the HKH region. Table 5.3 indicates the use of these other fuels in the region. Their use emerges mainly as the result of a decreasing supply of fuelwood, and depicts the stress under which forest resources are. The shift is also due to the low affordability of the mountain people (see also Figure 5.1). The transition from biomass to commercial energy forms is at a slow pace in the region due to price and nonprice factors as well as nonsuitability of technological options and lack of appropriate forest management.

Forest resources are being extracted far beyond their regenerative capacity in many parts of the HKH region. The few exceptions are isolated pockets such as eastern mountains of India and western part of Nepal, and places where accessibility limits the extraction of fuelwood and timber. Availability of forest resources within the HKH region of China is high (1.2 ha per capita) but, because of the inconvenience of transportation, people tend to use fuelwood available locally. This has

Table 5.3. Population, forest area and consumption of biomass fuels in Hindu Kush Himalayan Region, 1994–95

			Per capita consumption of biomass		
Area/Location	Population (Million)	Forest Area (Mha)	Fuelwood (kg)	Agric. Residue (kg)	Animal dung (kg)
HKH, China	19.7	23.6	760	460	321
Eastern Mountains, India	14.7	15.2	758	135	73
Western Mountains, India	18.3	5.6	635	81	88
Eastern Nepal	11.8	2.7	565	47	97
Western Nepal	8.6	6.2	691	26	49
Northern Mountains, Pakistan	22.3	18.5	290	54	87
Western Mountains, Pakistan	6.7	34.7	607	5	87

Mha = Million hectare

Note: The eastern mountains, India includes Sikkim, Darjeeling, and the North East Region. The western mountains, India includes Uttaranchal, Himachal Pradesh, and Jammu & Kashmir. The Eastern Nepal includes the Eastern and Central regions. The Western Nepal includes the West, Mid-west and Far-west regions. The western mountains, Pakistan includes Balochistan. The northern mountains, Pakistan includes the North West Frontier Province, Federally Administered Tribal Areas, and Ajad Jammu and

resulted in overexploitation of nearby forests. If explored on a sustainable basis, the fuelwood from the accessible forest area meets only about 35 per cent of the total demand for fuelwood.

There is a substantial difference in recorded forest area and actual forest cover in the hills and mountains of India, whereas the difference between recorded and actual forest cover is only about 2 per cent if aggregated for the whole HKH region in the country (see Table 5.4). The per capita forest area available in the HKH region of India amounts to 0.6 ha, while it is 0.85 ha in the eastern part and 0.32 ha in the western part of the HKH. The situation differs widely if examined at district or state level. For example, the availability of forest area per capita in Arunachal Pradesh is 7.9 ha, 0.34 ha in Jammu and Kashmir and 0.38 ha in Uttarakhand, while it is 1.3 ha in Uttarkashi and 0.23 ha in Nainital district.

The per capita accessible forest area of Nepal amounts to 0.37 ha in the mountains, and 0.29 ha in the hills. In terms of mapped forest area (i.e. includes area for potential forest regeneration), 0.93 ha per capita is available in the mountains, and 0.58 ha in the hills. The estimated quantity of fuelwood supply that can be obtained on a sustainable basis in Nepal amounts to 7.5 million air-dried tons (i.e. 366 kg per capita), instead of the average fuelwood consumption of 640 kg per capita. However, the fuelwood balance at district level shows different patterns. Fuelwood is in surplus in some of the districts in Western Nepal, while it is in short supply in central hills.

Table 5.4. Total forest area in relation to land area (1993 assessment)

Description	Recorded forest area (km^2)	Actual forest cover (km^2)	Land area (km^2)	Forest area (%)	Actual observed area (%)
Western Mountains	92 139	55 603	345 216	27	16
Himachal Pradesh	37 591	12 502	55 673	68	22
Jammu & Kashmir	20 174	0.443	238 418	8	9
Uttaranchal	34 374	22 658	51 125	67	44
Eastern Mountains	120 017	151 881	219 748	55	69
Hills of West Bengal	1011	697	9376	11	7
Sikkim	2650	319	7026	38	44
Hills of Assam	9314	7433	26 701	35	28
Arunachal Pradesh	51 540	68 661	83 743	62	82
Manipur	15 154	17 621	22 327	68	79
Meghalaya	9496	15 768	22 429	42	70
Mizoram	15 935	18 697	21 081	76	89
Tripura	6292	5538	10 486	60	53
Total Mountains	212 156	207 484	564 946	38	37
All India	770 078	640 107	3 295 273	23	19

Source: Rijal, 1999.

Public forest in the mountain areas of Pakistan (76 per cent of total forest area in Pakistan) provides most fuelwood required for domestic and industrial purposes of the country. About 14 per cent of the Northern mountains of Pakistan are covered by forest, though a significant variation from 15 to 60 per cent in forest cover is observed, when analyzed at the district level. The total sustainable supply of fuelwood in Pakistan is less than 40 per cent of the total demand but, in the fuelwood supply and demand balance for the Northern mountains, supply exceeds demand by 1.6 times.

5.4. MAJOR ISSUES PERTAINING TO FUELWOOD

What type of energy use pattern or mix is environmentally safe, economically sound, and culturally inoffensive in the mountains is a major policy issue being faced by the planners as well as beneficiaries of forests in the HKH region. In this context, a number of issues need to be considered and dealt with when promoting a fuelwood program to achieve sustainability. These issues are (Rijal, 1996, 1999):

- Prevailing unsustainable trends of energy supply and demand;
- Nonharmonious energy transitions towards noncommercial, low quality energy forms and/or towards nonrenewable fossil fuels;

- Wrong choice of energy resources and technologies as a result of lack of perspective related to both quality and quantity of energy in program design;
- Ignorance of biophysical aspects of mountain areas;
- Weak gender participation in decision making;
- Lack of understanding of sociocultural issues;
- Lack of suitable institutional framework to promote decentralized renewable energy technologies; and
- Methodological dilemma to internalize environmental concerns.

Ignoring these issues have led and will continue to lead to environmental consequences. The impacts are not limited to the decrease in forest cover and quality in the hills and mountains but involves also decrease in soil fertility, agricultural productivity, and water availability in springs, and increase in soil erosion and landslides (Myint and Hofer, 1998). Some research studies have also indicated that the decrease in forest cover may have led to a series of environmental damage downstream (Durning, 1993; Shen and Contreras-Harmosilla, 1995; RWEDP, 1993; Rijal, 1996).

Improved accessibility in some mountain areas through the construction of physical infrastructure has led to better living conditions with improved social infrastructure. However, this has come at the cost of encroachment around productive forests and their utilization to meet the timber and fuelwood requirements of the plains. Initially, mountain people were not aware of this situation and the cash flow was welcome. Many mountain settlers moved out to exploit better opportunities in the plains, and this hampered the diversification of mountain economies. Inappropriate forest policies applied in the mountains led to further dismantling of traditional and indigenous practices of managing forest resources (FAO, 1993).

5.5. FUTURE DIRECTIONS FOR WOOD ENERGY DEVELOPMENT IN THE HKH REGION

In the past, most governments and donor agencies considered fuelwood as a mere energy demand and supply problem (RWEDP, 1993). The diagnosis of the problems and design for solutions have been based on simple models of supply and demand, i.e. gap theory (Soussan, 1993). This has led to programs for planting trees, reducing consumption through the introduction of improved cooking stoves, and upgrading of the quality of biomass fuels.

The approach adopted for the dissemination of improved cooking stove (ICS) was essentially technology-focused, and with a few notable exceptions in the plains, these efforts have failed to have lasting impact on fuelwood scarcity. The failure of such technology-focused attempts has been documented in Rijal (1996) and RWEDP (1997). These interventions ignored the multiplicity of existing traditional technologies and disregarded sociocultural values. Since consumption of fuelwood was perceived as the main cause of deforestation, other factors such as collection of fodder for livestock, land needs for cultivation, and large-scale felling of timber, never received appropriate attention. Alleviation of human drudgery, improvement of deteriorating health conditions and the problems of soil fertility were never considered seriously.

The frequent failure of many such policy initiatives has led to a reappraisal of the fuelwood crisis. A number of studies (Soussan, 1993; RWEDP and ICIMOD, 1997) argue that biomass fuel production and use are intimately integrated into broader processes of resource management in local production systems. Fuelwood problems are likely to emerge gradually, as people respond to a variety of resource stresses. This means that fuelwood stress rarely manifests itself as a simple shortage of fuel (Soussan, 1993).

The issues of control over decisions concerning land and other resources are at the heart of effective fuelwood policies and programs in the context of mountain areas. In previous attempts, local and national governments have failed to establish the conditions that would allow efficient and sustainable allocation of land and resources for woods and cropland, or wood production and food (Durning, 1993; FAO, 1993; WWF and IUCN, 1996).

Although activities to be undertaken in any fuelwood strategy will vary according to prevailing local conditions in the mountain areas of each particular country, in general policy interventions should seek the following (Soussan, 1993; WWF and IUCN, 1996; RWEDP, 1997; Rijal, 1999):

- To secure property rights, and especially ensure the rights of those groups experiencing the worst problems over access to fuelwood resources. This should include customary and communal rights as well as private property rights.
- To improve market functions by eliminating policy-induced distortions in the prices of different types of energy resources and technologies. For example, decision makers are never concerned with the prevailing subsidy on commercial fuels (electricity and petroleum fuels) but always emphasize the commercialization of new and renewable energy technologies. All kinds of energy resources and technologies should be judged by providing a level playing field.
- To improve access to, and management of various renewable energy technologies and commercial fuels so that the options are made available for the mountain people to make appropriate decisions about their energy requirements.

- To bring the voice of the community to the fore, and build effective institutional structures to give the actors on the ground a real control over the decisions that affect their lives.

There is a need to develop specific fuelwood sector strategies that can capture the local specificity of fuelwood problems and opportunities in mountain areas. This should be accompanied by (Rijal, 1996; Soussan, 1993; RWEDP, 1997):

- Improvement on information, including the creation of a database at the lowest level of planning so as to understand the dynamics of biomass fuel production and use in mountain areas;
- Strengthening of the capability of fuelwood planning institutions at local and national levels to create an effective implementation capacity;
- Strengthening of coordination between different agencies as fuelwood issues are inherently cross-sectoral in nature; and
- Efforts to involve local people more (particularly women) in the planning and implementation of forestry and fuelwood programs.

Movement in this direction has begun in recent years, and a number of issues are emerging in the context of hill and mountain areas of Nepal. Some of these issues are also relevant in other countries of the HKH region. A few examples of critical issues are mentioned here.

- A community forestry program was being implemented in Nepal based on the demand of forestry users' groups. The program involved the transfer of forests to the users' groups and had a demand-driven approach. After the promulgation of the Forestry Act 1993 and the Forestry Regulation 1995, the Department of Forest rushed to formalize the forest users' groups and to hand over forest patches to the communities. This was done without a proper assessment of the wood energy needs of different groups. At this point, the program took a more conservationist approach focused on the formalization of property rights. The richer section of the population took control of the forest users' groups after they received legal status. As a result, poorer groups were given limited access to forest resources, which led to further marginalization.
- It is important to pay attention to the sustainability of supply and demand of fuelwood, including also other forest products such as medicinal herbs, fodder, and timber. This should be considered under various forest management types such as community forest, leasehold forest, social forestry, Joint Forest Management, and private forest. There are more than 200 community forestry users' groups and other types of forest management practices within a district

(1000–2000 km^2) of Nepal. Within a particular district some of the user groups are large (300 people) but have ownership of small patches of forest (10–15 ha). Meanwhile, some user groups are small (20–50 people) but own 100 ha or more. In the former case, women and vulnerable groups (low caste) have limited access and, in many instances, are bound to collect fuelwood from government forests for their own consumption, or to sell in the village or other markets and make their livelihoods. In the latter case, there is a surplus generation of cash income as the result of sales of forest produce, which has been the cause of many conflicts.

- Another important issue that needs consideration refers to the costs and benefits of forest management interventions (including clean energy development, community forestry and watershed management) and related environmental services. While such costs and benefits are envisaged within the boundaries of the upstream areas and communities, most of the benefits would accrue to the communities living in the adjoining plains or valleys. What kind of policies and institutional arrangements would allow transfer of these benefits to the upstream communities, so that the cost they bear for interventions is reduced? There is need to develop methodologies that better assess and help to distribute the social costs and benefits of development interventions in upstream and downstream communities.

The aforementioned issues need to be carefully examined. Central to these issues is the possibility to capture the diversity and dynamism of local fuelwood and energy situation in the mountains in a broader context of community development. This is a key to making program interventions relevant and responsive to the needs of the mountain people rather than let them become hostages of any particular set of vested interests.

A broader framework for the management of forest resources in mountain areas can help meet daily energy needs better while also reducing the rate of deforestation, reducing the loss of life due to land slides, increasing the crop productivity by reducing soil erosion and loss of soil nutrient, and balancing the seasonal fluctuation of water levels in the streams and rivers originating from mountains. In addition, such a framework shall also help to reduce flooding of the adjoining plains, regulating the flow of water and timber for people residing in the plain areas.

REFERENCES

Durning, A.T. (1993) *Saving the Forests: What will it Take?* Worldwatch Institute, Washington DC.

FAO 1993 (1993) Forestry Policies of Selected Countries in Asia and the Pacific, FAO, Rome.

FAO-RWEDP (1997) Regional Study on Wood Energy: Today and Tomorrow in Asia, FAO, Bangkok.

ICIMOD (International Centre for Integrated Mountain Development) and CRT (Centre for Rural Technology) (1997) *Manual of Rural Technology with Implications for Mountain Tourism*, Kathmandu.

Myint, A.K. & Hofer, T. (1998) Forestry and Key Asian Watersheds, International Centre for Integrated Mountain Development (ICIMOD), Kathmandu.

Ramana, P.V. (1998) As if Institutions Matter: An Assessment of Renewable Energy Technologies in Rural India, Technology and Development Group, University of Twente, The Netherlands.

Rijal, K. (1991) Technology Assessment, Planning and Modelling of Rural/Decentralized Energy Systems in Nepal, PhD dissertation, in Centre for Energy Studies, Indian Institute of Technology, Delhi.

Rijal, K. (1996) *Developing Energy Options for the Hindu Kush – Himalayas: Rethinking the Mountain Energy Development Paradigm*, ICIMOD, Kathmandu.

Rijal, K. Ed. (1999) *Energy Use in Mountain Areas: Trends and Patterns in China, India, Nepal and Pakistan*, ICIMOD, Kathmandu.

Rijal, K. (2000) Energy From the Hindu Kush – Himalayan Mountain Forests, in *Forests in Sustainable Mountain Development: A State of Knowledge Report for 2000*, Eds. Price, M.F. & Butt, N., CABI Publishing in association with The International Union of Forestry Research Organizations (IUFRO), U.K.

Rijal, K. (2001) Sustainable fuelwood use in mountain areas, in *Mountains of the World: Mountains, Energy and Transport*, prepared for The UN Commission on Sustainable Development (CSD) 2001 Spring Session, Mountain Agenda, Swiss Agency for Development and Cooperation (SDC), University of Berne, Switzerland.

Shen, S. & Contreras-Hermosilla, A. (1995) Environmental and economic issues in forestry: selected case studies in Asia, The World Bank, Washington DC.

Soussan, J. (1993) Relating macro-economic and sectoral policies to wood energy supply and use, in *Wood Energy Development: Planning, Policies and Strategies*, Ed. RWEDP, **Vol. III**, RWEDP, Bangkok.

WWF and IUCN (1996) Forests For Life, WWF International and IUCN, WWF-UK.

Chapter 6
Modernizing Cane Production to Enhance the Biomass Base in Brazil

Oscar Braunbeck, Isaías Macedo and Luís A.B. Cortez

6.1. BIOMASS AVAILABILITY CAN BE ENHANCED IN BRAZIL

The most important biomass sources in Brazil are sugarcane and forest residues. The Brazilian sugar industry is almost as old as the country itself. It was based on traditional production systems for many centuries but had a turning point in 1930 when president Vargas created the Sugar and Alcohol Institute (IAA). The earliest experiments utilizing cane ethanol date from that period. However, radical changes would not take place until the Brazilian Alcohol Program (PROALCOOL) was created in the 1970s, leading to a significant expansion of sugarcane plantations in Brazil.

Today, Brazil is the largest sugarcane producer in the world, being responsible for nearly 25 per cent of the total cane production, 13.5 per cent of the sugar production and 55 per cent of the world's ethanol. The cultivated area covered by sugarcane plantations reaches more than 5 million ha or 1.5 per cent of the total arable land in the country. Sugarcane production reached 340 million tons of cane in the 2003/04 season resulting in 24 million tons of sugar and 14 billion liters of ethanol.

The ethanol industry provides fuel for approximately four million cars driven exclusively with ethanol and approximately 24 per cent of the fuel being used in the rest of the country's car fleet. The sugar and ethanol industry generates a turn over of US$ 12 billion and creates 600 000 direct jobs in activities from agriculture to industry. It is a sector almost entirely owned by local entrepreneurs that has a significant potential to increase its participation in the country's economy through a more intensive use of by-products.

Traditionally, sugarcane has been harvested by hand requiring the elimination of leaves by combustion in the fields. The cane-burning technique destroys nearly 25 to 30 per cent of the energy potential in the cane, which is a strong drawback seen from the perspective of the surplus energy that can be generated. However, environmental laws are being issued restricting cane burning, especially near urban areas, which is paving the way for green cane harvesting practices. By green cane harvesting, we

Bioenergy – Realizing the Potential

mean that the burning process is eliminated allowing for the full utilization of the cane biomass material. Green cane harvesting will require a significant move towards mechanization, a process that has only started in the Brazilian sugarcane sector.

In this chapter, we evaluate the technologies available for green cane harvesting, how they need to be improved, and how they can help enhance the biomass available for energy conversion in Brazil. We analyze some of the difficulties and barriers that need to be addressed in favor of the adoption of such technologies. We focus particularly on the technological barriers, showing how productivity of sugarcane production systems can be improved, while also helping improve the overall economy of this industry.

6.2. THE SUGARCANE INDUSTRY AS ENERGY PRODUCER

Sugarcane production in Brazil increased from 224 million tons in 1989 to 300 million tons in 1998 and 340 million tons of cane in 2003/04. The fraction of sugarcane used for ethanol production was close to 50 per cent in 1998 as opposed to approximately 65 per cent in 1989. Thus the ratio of ethanol to sugar has decreased, but still half of the cane is being used for ethanol production.

As shown in Table 6.1, the average energy balance (output renewable energy/fossil input ratio) in the Brazilian ethanol production is 9.2, an exceptional figure also when compared with other biomass systems (i.e. ethanol from corn in the US). It should be kept in mind that the figures provided in Table 6.1 are based on the current average situation of sugarcane mills in Brazil. The majority of them use bagasse to meet their needs for electric power and thermal energy, but utilize low efficiency steam-based cogeneration systems for this purpose (Goldemberg et al., 1993). As will be gathered from the discussions here, there are opportunities to further improve this balance.

Table 6.1. Average input–output energy flows for (burned) cane production and sugar mill with ethanol distillery, in São Paulo, 1996 (in MJ/t cane)

	Input of fossil energy	Energy output
Sugarcane production and delivery (agriculture)	190	
Cane processing (industry)	46	
Ethanol Produced		1996
Bagasse Surplus		175
Total (external flows)	236	2171

Source: Macedo (1998)

Table 6.2. Estimates of bagasse and trash availability for energy generation

Trash recovery percentage	Trash recovered t (DM)/t cane	Total bagasse t (DM)/t cane	Total biomass available t(DM)/t cane	Fuel oil equivalent kg/t cane
100	0.075	0.14	0.215	77
50	0.0375	0.14	0.177	61

Obs: Equivalencies include fuel processing losses, storage losses, combustion expected efficiencies.
DM = Dry matter
Source: Copersucar (1998)

In 1998, the sugarcane industry contributed nearly 200 000 barrels of oil equivalent of ethanol per day to the Brazilian energy system. This contrasts with the Brazilian domestic production of nearly 1.4 million barrels of oil per day. In addition, the bagasse was being used in the cogeneration of electric, mechanical and thermal power. Bagasse is used at a rate of 0.14 ton dry matter per ton of cane, leading to an annual production of 40 million tons (dry matter). Nearly all the power from bagasse is used in-house at the mills. Estimates indicate a use of 3600 GWh for electric power plus 4500 GWh for mechanical drives, in addition to all the thermal energy requirements for processing sugarcane to ethanol and sugar.

A recent legislation is the restriction of sugarcane burning in the largest production areas of Brazil. It is expected that at least 50 per cent of the cane in the major producing areas will be harvested without burning and with mechanized technologies in the next 10 years. This shall result in a substantial increase in biomass availability. Together with more efficient technologies, this shift can significantly improve the overall energy balance of the system as a whole.

Table 6.2 shows an evaluation of the biomass that can be made available for energy generation as the new practice is established. The figures are based on an average trash availability of 10 tons dry matter/ha when approximately 55 per cent of the total planted area is harvested without burning the cane (55 per cent is an approximate value for the sugarcane area which can be harvested mechanically with today's technology). Table 6.2 provides calculations for 50 and 100 per cent trash recovery since this may vary depending on agronomic considerations. The following variables have been accounted for:

- Average amount of trash in sugarcane (tops and leaves);
- Main agronomic routes to harvest green cane with trash recovery;
- Soil properties with trash left on field;
- Advantages of trash left in field for herbicide elimination;
- Trash properties as fuel;
- Recovery costs;

- Utilization of biomass as boiler fuel or for gasification;
- Environmental impacts of trash recovery/utilization.

For an annual yield of 300 million tons of cane, with a trash recovery of only 50 per cent in 55 per cent of the planted area, the annual amount of new biomass available will be around 11 million tons (DM). This biomass can be effectively used for the purpose of energy generation. Conventional technologies can be employed, such as traditional biomass fired steam boilers and furnaces. Other technologies are also being seriously considered, particularly gasification and power generation with gas turbines, and hydrolysis followed by fermentation to produce ethanol. Pilot plant developments in both areas are underway.

However, it is important to point out that the potential utilization is attractive even with conventional technologies. The development of the cogeneration energy market in Brazil is overdue. The privatization of the electricity markets, the energy crisis of 2001 and the Kyoto Protocol to the Climate Convention have raised the interest for these new opportunities which can be gradually realized as the Brazilian economy recovers. Yet, the future development is not to be taken for granted. Gas markets based on imported gas from Bolivia shall become a strong competitor in many applications (see also Walter et al., Chapter 9). Thus, there is a risk that the better environmental choice based on domestic biomass resources is left behind.

6.3. RESEARCH AND TECHNOLOGY DEVELOPMENT IN SUGARCANE AGRICULTURE

Sugarcane is planted in Brazil mainly for sugar and ethanol production. To meet the requirements of production, the cane is cultivated utilizing the so-called ratoon system in which the first cut is made 18 months after plantation, followed by annual cuts along a period of 4 or 5 years, with decreasing yields. The Brazilian warm climate with rainy summers and clear skies in the winter help the cane to build a strong fiber structure during its growth phase and fix sugar in the winter.

The development of sugarcane crops in Brazil has benefited from constant research in the development of new varieties, particularly by the Agronomic Institute of Campinas (IAC), Cooperative of the Sugar and Ethanol Producers of the State of São Paulo (Copersucar) and the National Plan for Sugarcane (Planalsucar), a division of the Sugar and Alcohol Institute (IAA). In the last 30 years, however, the entire agricultural research apparatus of São Paulo, including research stations, were submitted to constant dismantling, which slowed down productivity improvements in sugarcane production at a time when oil prices were falling. This made ethanol less competitive. Meanwhile, the Federal Government pursued policies to eliminate state intervention in the sugar business, extinguishing the IAA and the Planalsucar, among other organizations.

Some of the earlier research responsibilities have been taken over by private organizations. Copersucar, for example, shifted the emphasis of its program to develop new varieties mainly through CTC, a technology center in Piracicaba, São Paulo. Another constant preoccupation in research has been related to extending the crushing season by developing early ripening, and increasing the yield by combating pests. The sugarcane breeding program has also incorporated modern techniques in molecular biology. Genetic transformation of sugarcane varieties has been achieved in the CTC laboratories and various transgenic sugarcane varieties are being presently field-tested.

Another major technological change observed in the sugarcane agriculture in the last 40 years is related to the introduction of machinery in soil preparation and conservation, particularly in the Southern states of Brazil. Some intensification of machinery use has been observed in the last two decades in various operations, from soil tillage to harvesting and, particularly, in cane loading. Less significant advancements were verified in planting the cane. All the cane is still being manually planted, although a cane-planting machine has been recently developed by Copersucar and licensed to DMB. Sermag and Brastoft are also offering planting machines in the market.

Despite the progress achieved in cane production so far, harvesting remains the least advanced operation. Cane fields are now systematically burned to allow manual harvesting. However, this is changing rapidly. Environmental pressures, legislation enforcement and cost reduction are pushing for mechanical harvesting of unburned cane (Furlani et al., 1996). In addition, the potential to generate revenues from cane residues is likely to provide incentives in this direction once the producers start evaluating other revenue options in their total production chain.

Development in cane harvesting and their potential to reduce agriculture costs will be discussed later in this chapter. Other possibilities for cost reduction are related to optimization in agricultural management, including introduction of operation research techniques and precision agriculture, which will allow a more rational use of resources and increase of yields. For example, the concept of environments of production combines soil charts with climatic and variety data and brings returns around US$ 40/ha according to tests conducted in some sugar factories.

Information software provides the basis for significant improvements in agriculture too. Software based on GIS (geographical information system) with embodied electronics together with productivity data can indicate the effect of productivity related variables such as soil fertility, pests, diseases, insects, weeds, soil compaction, and harvesting methods. This may serve as the basis for better management of crops and improved productivity. Software to improve logistics are also being developed, helping to improve the allocation of loaders, harvesters or trucks to optimize raw material flows to the factory. Logistic optimization in cutting and transporting cane has already brought reductions of 5 to 13 per cent in the agricultural cost of cane (data based on Copersucar mills in São Paulo, close to 25 per cent of Brazilian production in 1999).

6.4. FROM CANE BURNING TO MECHANICAL HARVESTING

At present, there is no single mechanical harvesting system available to handle the wide range of field conditions prevailing all over the world. Field conditions can vary from hilly land and presence of rocks to dry or wet soils requiring planting either at the bottom of the furrow or at the top of the ridge. Cane yield can vary from 60 to over 200 t/ha, the stalks being erect or recumbent with length in the range of 2 to 5 meters. Since the Brazilian landscape is hilly, planting and harvesting methods are different from those used in flat regions.

The Australian chopped cane principle and Louisiana's whole cane system are the two main harvesting technologies available in the world today. Other cane producers that are performing mechanical harvesting use derivations of these systems. Cuba, for example, utilizes KTP cane harvesters, which have a design based on the Australian technology principle (Gómez Ruiz, 1992).

The development of Australian technology was motivated by lack of labor for cane harvesting. Prototypes designed by farmers in the 1960s introduced chopping as a way to mechanically transfer the harvested product by free fall to the transport vehicle running side by side with the harvester. Chopping eliminates the whole cane loading operation. However, the cost of an adequately managed whole cane harvesting system can be inferior to that of a chopped one if cane losses and harvester idle time are accounted for in the cost analysis (Braunbeck and Nuñez, 1986).

Louisiana's *soldair* is economically the most efficient mechanical harvesting system for whole cane available at present (Richard et al., 1995). This harvesting system was developed for erect cane, which usually have short growing periods of about 7 months. As a result, it is not satisfactory to cut and feed the Brazilian cane crops, mainly the first cut, since the cane can easily fall at the harvesting period.

In the US, four different sugarcane regions utilize different harvesting practices. In Texas, the chopped cane system is used (Rozeff, 1980). In Hawaii, a locally adapted *push-rake* system is used which combines higher harvesting costs and lower cane quality. In Florida, the various systems described so far are found, that is, chopped cane, whole cane and manual cut.

In Colombia, where the government has set the year 2005 as the deadline to eliminate preharvest burning, work is underway to develop a harvester for collecting and chopping cane and cane residues. A major challenge in the development of the machine is to reach the high yields that characterize Colombia's Valle del Cauca sugarcane producing region (Ripoli et al., 1992).

Changing from burning to a full green cane harvesting practice requires comprehensive planning at various levels. The mechanical systems mentioned earlier are not optimally suited for Brazilian green cane either from the standpoint of harvesting costs, or from the point of view of the cane quality or losses incurred in harvesting. Technical adaptations are required to meet topographic, agronomic and sugarcane processing needs typical to the Brazilian context. Low-cost appropriated technology is still required to overcome not only cane harvesting difficulties but also trash recovery, baling and transportation. The full implementation of these technological shifts will only be possible if an innovative generation of engineers is able to eliminate the bottlenecks prevailing from cane mechanization to trash recovery and utilization.

6.5. TOWARDS MECHANIZED GREEN CANE HARVESTING IN BRAZIL

The practice of clearing and burning the soil has been used in Brazil since colonial times. Also in sugarcane production, burning has been commonly used as it increases the throughput in both manual and mechanical harvesting. Producers utilizing green cane harvest practices report 30 to 40 per cent lower daily tonnage for unburned cane when compared to burned cane (Ripoli et al., 1990). The sugarcane burning practice is now undergoing severe restriction due to increasing urbanization, particularly in southern parts of the country. Still, in the State of São Paulo, green cane harvesting is only practised in a 1 km radius around the cities as a result of law enforcement (Governo do Estado de São Paulo, 1988).

The reason for preventing the burning of sugarcane fields is to avoid emissions to the environment (i.e. pollutants such as CO and particulates), which have impact on human health (i.e. respiratory illnesses) and human amenity. Furthermore, valuable cane residues that could otherwise be utilized for energy purposes are lost in the burning process. Long-term trash mulching can reduce nitrate fertilizer applications by 40 kg N/ha, mainly as a result of reduced nitrate leaching (Vallis et al., 1996).

Field burning also results in sucrose losses by exudation on the surface of cane stalks. Ripoli et al. (1996) have found ethanol losses in the range of 59 to 135 liters/ha due to such practices. Work done by Fernandes and Irvine (1986) on commercial sugarcane fields of several companies indicated that the actual sugar yield was below the potential existing in the field, both through manual cut (−17 per cent) and through chopper harvester (−21 per cent). These losses occurred in burning, harvesting, loading, transportation, and reduction in cane quality.

The interest in mechanical harvesting grew strong in the 1970s due to studies that forecasted labor shortage (Stupiello and Fernandes, 1984). Mechanization efforts did not succeed at that time and the interest eventually faded in the 1980s, partly due to the deterioration of the Brazilian economy. By the middle of the 1990s, the question had regained interest (Furnari Neto et al., 1996). The main difference in studies done today is the emphasis on reduction of production costs, mechanical harvesting being one step in this direction. Today's cost for manual harvesting and loading of burned cane may exceed US$ 4.00/ton (Coletti, 1997) while mechanical harvesting hardly reaches US$ 2.00/ton (Lima, 1998)[1]. In the case of green cane harvesting, data is still unreliable but there are indications that manual cutting exceeds US$ 6.00/ton while mechanical harvesting is around US$ 3.00/ton.

Both the ergonometric and economic arguments indicate that green cane harvesting is likely to foster mechanization of harvesting practices in Brazil. A significant expansion of mechanized practices, however, will depend on improvements to the available mechanical harvesting technology. This includes the considerations listed below.

- The machine throughput and harvesting costs should not be but marginally affected by the amount of trash.
- It should be possible to remove the trash from the field for other purposes, such as energy conversion. So far, there are no adequate cane varieties and agronomic experiences to manage trash blanketed cane fields (Sizuo and Arizono, 1987).
- A percentage of the straw should be left in the fields for weed control and moisture conservation in cases where agronomic management techniques are well established.
- The present system of whole cane harvesting should be maintained to avoid unnecessary investments associated with the change to chopped cane as well as to avoid raw material (sucrose) loss associated with the chopping and cleaning processes, which represents an unacceptable technological step back in the Brazilian context.

The present field experience with whole cane harvesters together with recent developments on whole cane mechanical cleaning (Tanaka, 1996) as well as on machine right angle turning and pilling (Braunbeck and Magalhães, 1996) added to the known potential of computerized engineering resources applied to machine design, anticipate the feasibility to develop a whole cane harvesting equipment taking into account the

[1] These values refer to a period when the exchange rate of the Brazilian currency was temporarily high. At 2002 rates, the costs are significantly lower.

aforementioned features. Meanwhile, the mechanical harvesting expansion faces financial and technical constraints such as: shortage of skilled labor; bad field layouts and poorly performing harvester technology for the existing fields; lack of capital; existing whole cane transportation and reception at the factory different from the emerging chopped cane system; and design for maximum 12 per cent soil slopes for present harvesters which limits its use to about 45 per cent of the sugarcane areas.

Despite the country's large production of sugarcane, mechanical harvesting is still hardly employed in Brazil (see also Table 6.3). Although there is no precise figure on the number of harvesting machines in operation today, this number should not be greater than 600 machines harvesting approximately 50 000 tons/machine-season making a total of 30×10^6 tons. This represents 10 per cent of the total 300×10^6 tons harvested in the 1997/98 season. Frequent reference is made to the existence of quite high mechanization, but that relates to isolated cases such as Usina São Martinho, where 89 per cent of the cane was mechanically harvested already in the 1993/94 season.

Nevertheless, mechanical harvesting is growing fast in the State of São Paulo. This is not only due to the good topography of the state but also due to the well-developed road network and availability of skilled labor to operate a mechanized system. When cane producers of other states such as Goiás and Mato Grosso implemented mechanical harvesting, they faced serious difficulties in hiring adequate labor to operate and maintain the machinery. It takes several years until the harvester's fleet can reach a production of 400 t/machine-day as the season average in 24 h/day operations. Usina São Martinho, State of São Paulo, has exceeded productions of 600 t/machine-day, greatly due to skilled labor and adequate infrastructure, while Usina Sta.Helena, State of Goiás, achieved the average harvester throughput of 400 t/machine-day first after 6 years operation.

Chopped cane has higher dirt and trash content. Unloading of chopped cane frequently has lower priority at the mill as a function of its lower quality. It creates a transportation shortage to the harvesting fleet and increases the cost of the operation. Lower cane quality comes mainly from inadequate technology at the machine base cutter which consists of two flat disks with approximately 900-mm diameter each. This defines a 1800 mm wide plane and requires a perfectly levelled soil for the disks to operate very close to the surface without cutting the soil. Recommendations for efficient operation of base cutters require flat and levelled land, but this turns out to be insufficient since the problem remained in Australia even after many years (Ridge and Dick, 1988).

Brazilian undulated cane areas inject large quantities of soil into the harvester. Though it is removed inside the machine, about 0.5 per cent of the soil still remains in the cane and this has an impact in the cane processing at the factory. The soil

Table 6.3. Comparative features of sugarcane harvesting technologies used in Brazil

Parameter	Type of Harvesting		
	Semimechanized	Mechanized – chopped cane	Mechanized – whole cane
System features	Hand cutting with mechanical grab loading	Stalk and top cutting with simultaneous cleaning and loading	Stalk and tops cutting with cane bundling
Fraction of today's harvest	About 80% and decreasing	<20% (increasing)	<1%
Harvesting capacity	*Cutting:* 4–7 t/man-day; loading at 400 t/machine-day	4000 t/machine-day; loading may achieve 600 t/machine-day	600 t/machine-day; loading may achieve 700 t/machine-day
Cost	US$ 3 to 4/t (loading and overall labor costs)	US$ 2/t	US$ 1.5/t
Main restrictions	Lack of labor requiring labor import from other states; interruption of production because of regional strikes; need for training to maintain quality and productivity; not a viable alternative for green cane harvesting	Loss of raw material from base cutter, conveyor rolls, chopper and extractors intense traffic between lines; two passes by the harvester and by the following vehicle; overload of decanters at the factory; high sugar losses during washing	Requires trained labor; losses of raw material originated from base cutting and elevating rolls; traffic between lines; two passes by the harvesters; damaged stalks by the base cutter and transporting rolls
Advantages	Quality improvement; less raw material losses; avoids the set up of operation and maintenance	Reduced labor comparing with hand cutting; reduced harvesting costs; easy harvesting operations	Minimum incidence of labor (only in operation and maintenance); independence of cutting and transporting operations which eases the operation management; increases

	infrastructure of harvesters and specialized operation teams		productivity of cutting and transporting operations
Operating principle	*Cutting:* the cane stalks are cut at the base and deposited on 5-row mowing oriented towards the planting lines	*Cutting and loading:* double horizontal rotating discs for cutting helped by helicoidal rotating cones feeders (helpful for nonerect cane harvesting)	*Cutting and feeding:* double horizontal disks for base cutting helped by a pair of rotating cones with helicoidal edges for separating, elevating, and feeding the nonerect cane with intercrossed stalks between the lines
	Loading: grab loaders with hydraulic handlers mounted on tractors remove 600 to 1200 kg bunch that are transferred to the transporting vehicle (cost of loading: US$ 0.5 to 0.6/t)	*Tops cutting:* cutting by inertial disc with peripheral trapezoidal blades fed by two converging counterwise rotors (mainly for erect cane)	*Tops cutting:* inertial cutting through a disk with peripheral trapezoidal knives fed by two converging counterwise rotors
		Elevation and dirt separation: cascade of horizontal axes rolls with paired-mounted rotating in opposite directions with increasing tangential speed to the chopper	*Elevation and dirt separation:* cascade of horizontal axes rolls with paired-mounted rotating in opposite directions with increasing tangential velocity to the discharge
		Chopping and ventilation: two rotating axial knives with contrary and synchronized rotation chops and unloads by material in a pneumatic cleaning system to separate the leaves by terminal velocity	*Mowing and discharge:* the whole stalks are launched to the interior of a bin where they are accumulated to form a bunch to be discharged in regular intervals forming mows perpendicular to the planting lines. The traffic during the loading operation should be perpendicular to the furrows which has been a rejecting factor by the harvesters/users because of truck and loader overloadings

content of hand cut cane will not exceed 0.15 per cent when loaded with properly operated rotary push-pilers. Summarizing, the chopper harvesters present several critical points of cane losses such as the double disk base cutters, the feed rollers, the chopper and the cleaning extractors. These components are responsible for losses ranging from 7 to 15 per cent (Ridge et al., 1984; Fernandes and Irvine, 1986). The solution to this problem will arise from existing and future developments on alternative cutting and feeding mechanisms more than from insisting on extension work to further level cane fields, which creates adverse agronomic conditions for cane longevity and moisture conservation.

The technological innovation and investment capacity of the Brazilian agricultural machinery industry is rather small. It essentially limits itself to promoting the chopped cane technology developed abroad. There is also a technical barrier related to the potential use of machinery in areas with an unfavorable topography. Techniques, such as four-wheel steering and traction as well as steel or rubber tracks would allow to extend mechanization up to about 90 per cent of the areas which are currently occupied by sugarcane. Today's harvesters are mainly one-row machines with high center of gravity using two-wheel traction in the rear and two-wheel steering in the front. This driving and traction configuration is acceptable for agricultural tractors in which alignment with the line of motion is not so important. In the case of harvesters, the lack of machine alignment with the crop lines leads to cane losses and frequent stops due to clogging. It would be technically possible to develop harvesters capable of operating in most of the hilly areas if present harvesting principles were simplified to allow investment on four-wheel traction and steering, still keeping the equipment economically feasible.

A successful implementation of mechanical harvesting in Brazil needs to address a series of technical issues based on topographic, agronomic and sugarcane processing conditions, typical to Brazil. Therefore, a development and economic effort is still required to improve the harvesting efficiency. Table 6.3 compares the features of sugarcane harvesting technologies being used in Brazil, which further illustrates the technological improvements needed.

6.6. TRASH AND BAGASSE – SAME SOURCE BUT DIFFERENT FEATURES

Trash and bagasse are two fibrous materials from sugarcane, each with its own specific characteristics. Although sugarcane trash and bagasse are fibers of the same origin, their physical and chemical characteristics may differ significantly. The sugarcane bagasse has smaller particle size because it has been milled in the juice extraction process. This results in finer particle size when compared with unprocessed trash (Olivares et al., 1998).

Of the two, bagasse is the more studied material, because it is essentially an *industrial residue* that has been used for many decades as a fuel in bagasse boilers. Although it is well known that biomass fuels should have lower moisture content, the drying of bagasse has not been considered a very profitable process, at least until recently. For this reason, bagasse has been used with its original moisture content of approximately 50 per cent wet basis.

Regarding particle size, bagasse is a very homogeneous material, at least when compared with trash. Therefore, bagasse burning is more predictable a process because the biomass is composed of fibers with more or less similar characteristics, like composition and size. Dirt is also a much easier problem to solve for bagasse than trash, at least in conventional low-pressure boilers (2.1 MPa).

There is no existing technology and experiments on bagasse pelletization and briquetting of sugarcane trash. Limited information is available on bagasse pelletization but commercial bagasse briquetting has not been reported in Brazil (Bezzon, 1994; Cortez and Silva, 1997). Some technical difficulties are associated with the long-term integrity of the bagasse briquette. The bagasse high moisture content, nearly 50 per cent w.b., is considered the most negative factor when briquetting is considered. Bezzon (1994) conducted experiments at UNICAMP heating up the bagasse up to 200–300°C before briquetting (1 cm diameter and 2 cm length). The applied pressure ranged from 20–25 MPa and yielded briquettes with densities from 1000 to 1240 kg/m^3. The results were promising but experiments with larger briquettes were not conducted. It is known that heating the briquette can melt lignin resulting in a fiber-binding material.

Trash is essentially an *agricultural residue* that is now being seriously considered for energy purposes. Tops and leaves are the main components of sugarcane trash and they are removed at two different stages of the harvesting process. Top removal is the very first operation executed on standing stalks, before base cutting. Very little information is available in Brazil concerning trash recovery and use. Very few experiments and data are reported about its characteristics and how this influences equipment design and operation. Usina Sta. Elisa in São Paulo has conducted some experiments on trash recovery in cooperation with Dupont and Class. Also experiments are currently being conducted by Copersucar on trash recovery and its use in boilers baling trash with the *Class and Case* machines, but no conclusive reports are available so far.

The present harvesting technology has two main drawbacks in relation to top recovery. The first one is related to the topper not being able to reach the tops of nonstanding stalks. Efforts are being made to remove tops using the extractors at the second cleaning stage of the chopper harvester. This operation is highly inefficient and top removal is directly related to cane losses. The second one arises from the fact that after cutting and chopping, a low surface density of tops is left on the field for future collections, using adapted hay technology. Raking and windrowing of tops over a bare soil surface results in high dirt contamination. Field experiments indicated soil contents of 20 kg per ton of trash recovered using hay equipment (Copersucar, 1997). Harvester design should include trash recovering from the beginning. Two pieces of existing technology need to be incorporated into a cane harvester to make trash recovering more efficient.

Tops are green, high moisture residues that require field natural drying to improve biomass quality. After harvesting the green cane, trash may be left on the soil to dry for a few days. When the trash is nearly dry, with approximately 30 per cent moisture content, it can be recovered. If left in the fields, it may increase the risk of fire or may slow down ratoon sprouting. Thus it is generally accepted that at least part of the trash should be recovered. The specialists' recommendations on how much trash should be recovered vary from 50 up to 90 per cent. It is believed that organic material left in the fields may bring some agronomic benefits helping to control weeds and increasing the long-term soil fertility. An experiment has been reported by Molina et al. (1995), using a roller type trash baler, which processed 5.7 t/h, recovering 83 per cent of trash with 30 per cent moisture content and obtaining low-density bales with 120 kg/m^3.

A series of operations are required to perform trash recovery starting with raking the trash into continuous windrows after natural drying in the field. In the sequence, the trash must be baled to make transportation economically feasible. The commercial balers will compact up to 150 and 200 kg/m^3 density. The final product is supposed to resist transportation and storage in adverse climatic conditions when stored in the field. The operating costs may be the determinant in making the trash recovery feasible. The costs reported by Molina et al. (1995) varied from US$ 7 to US$ 25.00/t, depending on local conditions such as topography, infrastructure and available technology. It is a general consensus, in Brazil, that it will be difficult to compete with bagasse, but not so difficult to compete with other alternative energy resources, such as natural gas from Bolivia which is being offered in the market at a cost between US$ 2 to 3 per MMBTU. The systems tested in the last 5 years by Copersucar include various alternatives and resulted in costs below US$ 1/MMBTU in at least two cases.

Particle size is certainly a major consideration with trash because it may affect the conversion residence time in the reactor. This may also be affected by the moisture content in the particles, some of which have more moisture than others. The heterogeneous characteristics of the trash are certainly a drawback, which requires a fuel preparation procedure. Dirt in the trash is another major problem since the dirt increases the ash content and may interfere with the ash melting point and formation of deposits in the heat exchanger walls when combustion reactions take place.

In short, the existing boilers installed in sugar factories throughout the world can handle and operate better using biomass similar to bagasse. Any large variation in particle size, moisture content and dirt content will negatively affect the reactor efficiency and operation. Most likely, the most appropriated procedure is to prepare the fuel to meet the equipment requirements or, if possible, design a reactor that can efficiently operate within a larger spectrum of fuel properties. At UNICAMP an unbaling system is being conceived to unbale and feed biomass (trash) into a boiler (see Figure 6.1). In this project, the aim is to develop a technology that allows a continuous supply of biomass up to the boiler distribution system.

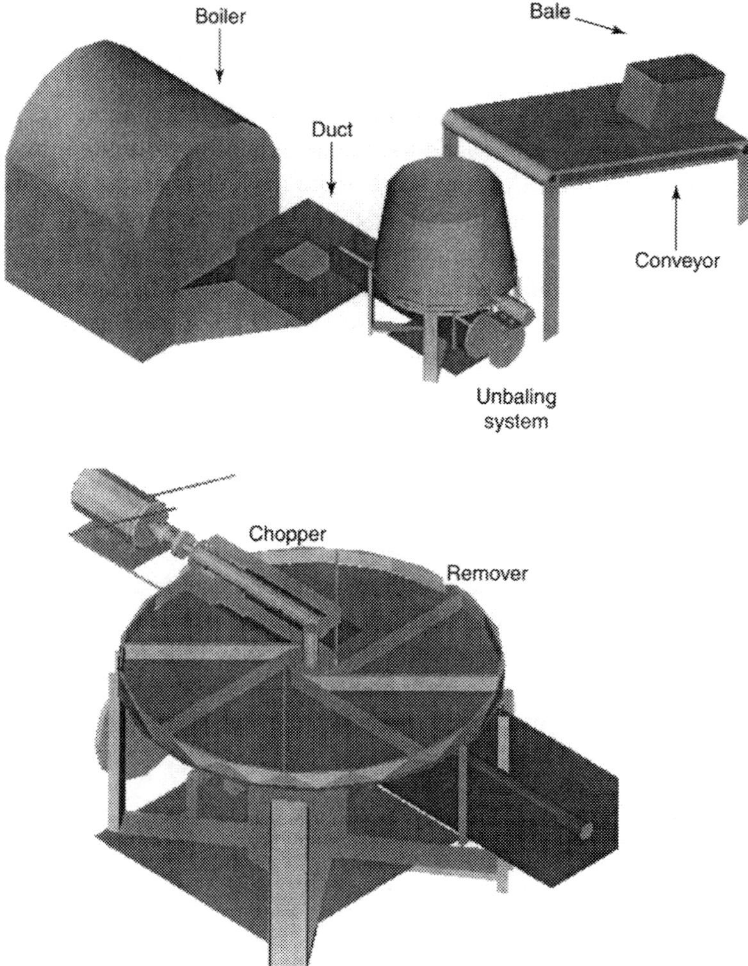

Figure 6.1. The UNICAMP unbaling-feeding system for biomass (trash).

The main task of the unbaling system is to cut the bale and then feed the biomass by means of the feeding screw. A chopper located at the silo's bottom performs the cutting. The remover helps by rotating the bale and placing it against the chopper. Figure 6.1 shows the unbaling system and the chopper. The system may also include a dryer to homogenize the material moisture content.

The costs involved in the biomass preparation are not negligible. Besides the necessary investments in equipment, there is also the need for capital for the system operation, particularly if a drying system is required and a storage facility is needed. This infrastructure and economics are being examined at UNICAMP and an in-factory system is being considered during the tests. An important drawback is that there is little information in the literature about large-scale biomass preparation and handling, except in forestry, and this information is essential in this kind of projects.

The economic use of trash will depend on investigating more cost-efficient technologies to handle, transport and use the material, transforming it into a more valuable commercial product. In this sense, more research is needed not only in the conversion for electricity, ethanol and fuel gas production but also in charcoal production. Charcoal has a well-established market in Brazil and its production is still based on very traditional technologies, based on low-efficiency furnaces and waste forest wood (Rosillo-Calle et al., 1996). The charcoal production from trash and bagasse could benefit both the sugar and the steel industries in Brazil, but no technology has yet been developed to adapt such by-products for this purpose.

6.7. USING TRASH AND BAGASSE FOR ENERGY PURPOSES IN DIFFERENT INDUSTRIES

From the energy point of view, the most important characteristics of a fuel are given by its composition, heating value and other properties related to the energy conversion technology where it will be used. Table 6.4 shows the results of an analysis conducted at the Alternative Fuels Laboratory at UNICAMP for eucalyptus, bagasse and trash from sugarcane. It turns out that the energy value of cane trash

Table 6.4. Composition and heating value of eucalyptus, trash, and bagasse from sugarcane

Sample	Moisture content (%)	Volatiles (%)	Fixed carbon (%)	Ash (%)	C (%)	H (%)	H.H.V. (kJ/kg)
Eucalyptus	11.9	80.2	19.8	0	49.6	6	18 494
Cane trash	10.5	74.7	15	10.3	43.2	5.6	15 203
Cane bagasse	9.9	75.4	10.8	10.8	43.6	6.2	17 876

was only 15 per cent lower than that of the bagasse at similar moisture content. Unfortunately, important characteristics such as the ash melting point, critical when a fuel is used in boilers, are not properly documented. Experiments on trash combustion are not reported either.

Wood from native forests has been historically used to provide useful energy through direct combustion. Also in Brazilian sugar mills, wood was and is still used to a limited extent to complement bagasse as energy source. Bagasse provides enough heat through direct combustion in steam boilers to meet the needs to crush the cane in the mills, to provide process heat in the factory operations and also to generate enough electricity by turbo-generators to drive all electric motors and meet other electricity requirements at the mill. Bagasse-based energy is not only enough to meet these needs but is also produced in excess in sugar mills and ethanol distilleries in Brazil. Many factories have to adjust their boilers to simply burn all excess bagasse because, otherwise, the material will deteriorate and pose risks.

Meanwhile, in the State of São Paulo, where most of Brazilian sugarcane is produced, fuel wood is used in the manufacture of bricks, in food industries, bakeries and restaurants. The wood is usually transported from distant locations using diesel trucks. It has become scarcer and more costly and is being gradually substituted for natural gas and even conventional burning oil.

There is a significant potential for bagasse to immediately substitute firing wood. The limiting factors are apparently associated with lack of entrepreneurship to disseminate a "*bagasse culture*" that helps promote a more diversified use of substitutes. One example of successful substitution is given by the Destilaria Rosa located in Boituva, São Paulo, where a small brick industry was installed just a few meters from the ethanol distillery. The excess bagasse, nearly 30 per cent of the total, is used to produce low cost bricks. Tests were conducted in their furnace to evaluate both efficiency and quality of the bricks (Aradas et al., 1998). In spite of good results, this example has not been emulated by other mills.

Bagasse is used in isolated cases in other industries, for example, in the vegetable oil industry. A more intensive use outside the sugar and ethanol sector has been observed in the orange juice industry, where bagasse boilers similar to those employed in the sugar mills are installed. The Brazilian orange juice industry is amongst the largest in the world and a large production is located in São Paulo, near the sugarcane production area. Unfortunately, the consumption of bagasse in the orange juice production does not create sufficient demand so as to provide a real incentive to the supply side. In fact, the demand for bagasse in the orange juice industry is presently threatened due to the availability of the Bolivian natural gas, which is leading to a review of bagasse contracts.

The market price for bagasse varies depending on the local availability and distance. Usually, bagasse is sold at prices ranging from US$ 5 to 12 per ton of bagasse (representing a range of US$ 0.60 to 1.35/MMBTU) in the core sugarcane regions in São Paulo. Small enterprises are producing equipment for direct combustion of vegetable residues, including cane bagasse. Such equipment is simple in conception and is being sold at prices around US$ 1600 for a feeding capacity of 500 kg of biomass/h.

6.8. REALIZING THE BIOMASS POTENTIAL IN THE SUGAR–ETHANOL SEGMENT

Cultivation of sugarcane in Brazil has experienced important growth after the implementation of the PROALCOOL program. Brazil is a major sugarcane producer, and the industry plays an important social and economic role in the country's economy. Together with recent developments, particularly regarding environmental legislation, sugarcane crops are about to undergo an important benchmark as green cane harvesting practices are disseminated.

The trash to be removed from the cane fields represent a significant surplus of raw material that can be utilized to the benefit of the sugarcane business total economy, possibly improving its competitiveness and attractiveness. In addition, green cane management may contribute to increase the amount of energy generated from domestic sources and to reduce CO_2 emissions at a global level. These factors enforce the importance to seriously consider a better utilization of the biomass generated from the sugarcane crops.

However, the green cane management is still new in Brazil. Many lessons remain to be learned about managing the excess biomass to be left in the fields together with the different practices to deal with pest controls caused by lack of burning. Complementary technology to collect, transport, and prepare the trash needs to be further developed particularly to reduce the most costly operations and counteract some of the most immediate technological drawbacks.

A crucial question to intensify biomass utilization in Brazil is how to reach competitive costs, guarantee regular supply, and commercially available technology. In terms of competitive costs, bagasse still has difficulties competing with fuel wood (US$ 3–5/ton) and other low-cost residues such as sawdust, cotton husks, coffee and peanut husks. In terms of supply, there is need to develop the supply chains so as to guarantee a regular functioning of biomass markets and conquer full credibility among users. By fully utilizing the energy potential of sugarcane crops, we will be improving the overall sustainability of the sugar–ethanol industry.

REFERENCES

Aradas, M.E.C., Cortez, L.A.B. & Silva, E.E.L. (1988) Bagasse as a brick kiln fuel in *International Sugar Journal*, **Vol. 100**(1189), pp 16–25.

Bezzon, G. (1994) Síntese de novos combustíveis sólidos a partir de resíduos agroflorestais e possíveis contribuições no cenário energético brasileiro, Master's thesis, Faculty of Mechanic Engineering, UNICAMP, Campinas, SP, Brazil.

Braunbeck, O.A. & Magalhães, P.S.G. (1996) Dispositivo virador de cana inteira, Instituto Nacional de Propriedade Industrial, UM-70 SP, Brazil.

Braunbeck, O.A. & Nuñez, G.J.S. (1986) Corte, carregamento e transporte – análise de custos e desempenhos, *Boletim Técnico Copersucar* N$^{\underline{o}}$ **35**/86, Centro de Tecnologia Copersucar, Piracicaba, SP, Brazil, pp 44–53.

Coletti, J.T. (1997) Principais procedimentos para redução de custos, in *Jornal Cana*, April, Brazil, pp 18–19.

COPERSUCAR (1997) Company information, Cooperativa de Produtores de Cana, Açúcar e Álcool do Estado de São Paulo Ltda, São Paulo, Brasil, p 121.

COPERSUCAR (1998) Geração de energia por biomassa: bagaço de cana-de-açúcar e resíduos, Projeto BRA/96/G31 (PNUD/MCT), Centro de Tecnologia Copersucar, RLP-04.

Cortez, L.A.B. & Silva-Lora, E. Eds. (1977) *Tecnologias de conversão energética da biomassa*, Editora da Universidade do Amazonas, Manaus, Brasil.

Fernandes, A.C. & Irvine, J.E. (1986) A comparison of the productivity of the chop-load system of harvesting sugarcane with the hand-cut, grab-load system in *STAB*, **Vol. 4**(6), Brazil, pp 105–110.

Furlani Neto, V.L., Ripoli, T.C. & Villa Nova, N.A. (1996) Avaliação de desempenho operacional de colhedora em canaviais com e sem queima prévia in *STAB*, **Vol. 15**(2), Brazil, pp 18–23.

Goldemberg, J., Monaco, L.C. & Macedo, I.C. (1993) The Brazilian fuel-alcohol program, in *Renewable Energy: Sources for Fuels and Electricity*, Eds., Johansson, T.B. et al., Island Press, Washington.

Gómez Ruiz, A. (1992) Sistema cubano de cosecha en Verde, Seminário TECNOCANA/92, Araras, SP, Brazil.

Governo do Estado de São Paulo (1988) Dispõe sobre a proibição das queimadas, Decreto No. 28.848, 30/08/88, São Paulo, SP, Brazil.

Macedo, I.C. (1996) Greenhouse gas emissions and energy balances in bio-ethanol production and utilization in Brazil in *Biomass & Bioenergy*, **Vol. 14**(1), Elsevier, pp 77–81.

Macedo, I.C. (1998) Greenhouse gases and bio-ethanol in Brazil in *International Sugar Journal*, **Vol. 100**(1189), pp 2–5.

Molina, Jr., W.F., Ripoli, T.C., Geraldi, R.N. & do Amaral, J.R. (1995) Aspectos econômicos e operacionais do enfardamento de resíduo de colheita de cana-de-açúcar para aproveitamento energético in *STAB*, **Vol. 13**(5), Brazil, pp 28–31.

Olivares Gómez, E., Brossard Perez, L.E., Cortez, L.A.B., Bauen, A. & Larson, D.L. (1998) Considerations about proximate analysis and particle size of sugarcane bagasse and trash, Paper # 98- 6011 of ASAE Annual International Meeting, Orlando, Florida.

Payne, J.H. (1991) Cogeneration in the cane sugar industry, *Sugar Series 12*, Elsevier Scientific Publishing Co., The Netherlands.

Richard, C., Jackson, W. & Waguespack, J.R. (1995) Improving the efficiency of the Louisiana cane harvesting system in *International Sugar Journal*, **Vol. 98**(1168), pp 158–162.

Ridge, D.R., Hurney, A.P. & Dick, R.G. (1984) Cane harvester efficiency, *Conference on Agricultural Engineering*, Bundaberg, Australia, pp 118–122.

Ridge, D.R. & Dick, R.G. (1988) Current research on green cane harvesting and dirt rejection by harvesters, *Proceedings of Australian Society of Sugar Cane Technologists*, pp 19–25.

Ripoli, T.C., Mialhe, L.G. & Brito, J.O. (1990) Queima de canavial – o despedício não mais admissível in *Álcool & Açúcar*, **Vol. 10**(54), pp 18–23.

Ripoli, T.C., Stupiello, J.P. & Martinho, W.L.R. (1992) A cana-de-açúcar no Valle del Cauca in *STAB*, **Vol. 10**(3), Brazil, pp 37–40.

Ripoli T.C., Stupiello, J.P., Caruso, J.G.B., Zotelli, H. & Amaral, J.R. (1996) Efeito da queima na exudação dos colmos: resultados preliminares in Anais do Congresso Nacional STAB, Maceió, Brazil, pp 63–70.

Rosillo-Calle, F., de Rezende, M.A.A., Furtado, P. & Hall, D. (1996) *The Charcoal Dilemma: Find Sustainable Solutions for Brazilian Economy*, Intermediate Technology Publications, London, U.K.

Rozeff, N. (1980) An investigation of sugarcane scrap in the Rio Grande Valley in *ISSCT*, **Vol. 1**, pp 22–27.

Sizuo, M. & Arizono, H. (1987) Avaliação de variedades pela capacidade de produção de biomassa e pelo valor energético in *STAB*, **Vol. 6**(2), Brazil, pp 39–46.

Stupiello, J.P. & Fernandes, A.C. (1984) Qualidade da matéria-prima proveniente das colhedoras de cana picada e seus efeitos na fabricação de álcool e açúcar in *STAB*, **Vol. 2**(4), Brazil, pp 45–49.

Tanaka, F. (1996) Limpeza de cana inteira, Master's thesis, Faculty of Agriculture Engineering, UNICAMP, Campinas, SP, Brazil.

Vallis, I., Parton, W.J., Keating, B.A. & Wood, A.W. (1996) Simulation of the effects of trash and N-fertilizer management on soil organic matter levels and yields of sugarcane in *Soil & Tillage Research*, **Vol. 38**(18), pp 115–132.

Chapter 7

Integrating Forestry and Energy Activities in Lithuania Using Swedish Know-how

Semida Silveira and Lars Andersson

7.1. BILATERAL COOPERATION FOR KNOW-HOW AND TECHNOLOGY TRANSFER

Biomass resources in the Baltic Sea Region are large, providing good ground for bioenergy. The interest to use local energy resources has increased significantly in the past few years, and many activities have taken place towards biomass utilization since the early 1990s. Nevertheless, significant amounts of forest residues are still unused and many opportunities remain to be explored. Lack of adequate logistic systems for harvesting, collecting and transporting biofuels constrains a broader use of these resources. Competition with other low-cost fuels and lack of supporting policies also hinder the development of bioenergy systems in the region.

Varied national regulations and taxation of fuels, lack of proper biofuel standards, limited financing opportunities for new projects, the need for upgraded infrastructure for logistics and new energy-related technologies are some of the barriers that need to be addressed to enhance the dynamics of bioenergy markets in the Baltic region. Meanwhile, sharpened environmental requirements, rising costs for imported fuels, and concerns about regional development and balance of trade are strong motives helping promote local fuels, thus opening a window of opportunity for biomass.

In fact, despite hindrances, increased trade activities with biofuels have been observed in the Baltic, and a number of bilateral and multilateral cooperation projects have been successfully carried out, emulating the experiences of neighboring countries. Since the efforts to increase biomass utilization have been particularly successful in Nordic countries, the accumulated know-how and experience is finding its way into the whole Baltic Sea Region. While initial efforts were particularly focused on biomass-based technologies for energy generation, the new steps have a broader focus and evaluate ways for an efficient organization of whole bioenergy systems at the regional level.

Bioenergy – Realizing the Potential

This chapter is based on a bilateral project developed between the Swedish Forest Administration and the Forest Department and the Ministry of Environment in Lithuania, with the support of the Swedish Energy Agency. The project has a starting point in the Lithuanian resource potential and institutional framework on the one hand, and the Swedish experiences with bioenergy systems, on the other hand. It looks at how the application of Swedish know-how in the form of mechanization and management practices can boost biofuel production in Lithuanian forests and help enhance bioenergy utilization in the country. A summary of the major issues assessed and evaluated are provided, indicating not only the complexity, but also the level of understanding and know-how accumulation that has been reached about biomass-based systems.

Initially, a feasibility study was carried out in the eastern part of Lithuania to identify conditions for the utilization of woodfuel within the seven state forest enterprises. The purpose was to find appropriate methods for profitable horizontal and vertical integration of the handling of forest fuels, and ways to integrate them into ordinary forestry and energy supply systems. The recommendations evolved into a demonstration project in Rokiskis state forest enterprise and capacity-building programs for continued cooperation and further development of the Lithuanian biomass potential.

7.2. FOREST MANAGEMENT IN LITHUANIA

Forests cover 31 per cent or almost 2 million hectares (ha) of the Lithuanian territory. Still the forest sector accounts for only some 3 per cent of the country's GDP. Lithuanian forests are characterized by a good variety of tree species, though pine, spruce and birch compose 80 per cent of the stands. The growing stock has more than doubled in the past few decades thanks to the strict control of fellings and expansion of forest areas. Today, the growing stock totals 371 million m^3, with an average of 193 m^3 per hectare, and annual increment of 6.3 m^3 per hectare. Over recent years, felling amounted to some 5 million m^3 annually. However, according to experts, an annual cut of 6.2 million m^3 or more could be maintained in the next ten years.

The low intensity of forest management practices in Lithuania has resulted in relatively dense forest stands. Consequently, the felling generally yields large volumes of firewood. Fuelwood is often handled in the same way as e.g. pulpwood, which implies considerable costs and reduces the overall profitability of forest management. There are, therefore, good reasons to improve this practice. Better management, including new practices for fuel collection and handling, could contribute to the overall improvement of the economy of forestry activities.

There is a large need for measures such as precommercial and commercial thinning, particularly as the volumes from clear cutting seem to be increasing. In 2001, the total cutting reached 5.7 million m^3, but Lithuania has enough wood resources and favorable age structure of stands to increase the wood supply significantly in the near future. If forest management practices are changed towards shorter rotation ages and more intensive precommercial and commercial thinning methods, significant expansion of the fuelwood supply can be achieved. In addition, improved economic use of abandoned agricultural land is possible. For a long time, former agricultural areas have been naturally afforested, mostly with nonindustrial species, but a better choice of species and new management practices can develop these areas for the production of woodfuel and industrial wood.

In Lithuania, fuelwood is mostly used for heating in households. Lately, biomass-based district heating systems have been developed. Sawmills have supplied most of the fuelwood used in district-heating plants, while the forest sector has played a minor role as supplier. Biofuel markets are still limited in the country, but are evolving, not least due to expanding import markets in other countries of the region, for example, Sweden. The structure of wood consumption is changing. Consumption of woodfuel increased more than twice in the household sector during the 1990s and is expected to increase further. The conversion of boilers, particularly for district heating, has been a major drive of this tendency.

The forest ownership structure has changed considerably after the restoration of independence in Lithuania in 1991. Since the beginning of the land reform, ownership rights were restored to more than 165 000 forest owners, who now control approximately 531 000 ha or 26 per cent of the total forest area. The process of forest restitution is still proceeding and it is expected that, after completion of the land reform, private owners will control about half of the total forest area. Private forests are usually small and the average area of a private forest holding is 4.4 ha.

Management of private forestland is a new phenomenon in Lithuania and so is forest owners' associations. There are two separate private forest owners' associations established in the country today but no more than one per cent of the private forest owners are members. The new forest owners need to comply with official policies for the sector but live under severe economic constraints. Associations can provide various services for members including information and consultancy, education in silviculture and forest management, and representation of the interests of private forest owners. Nevertheless, private forest owners are reluctant to join interest organizations because they do not seem to see the benefits. In particular, the idea of cooperatives do not seem very attractive to potential members.

The forest authorities in Lithuania have recently been restructured. At present, the Forest Department at the Ministry of Environment is the institution responsible for the Forest act and formulation of strategies, recommendations and guidelines to forest-related activities. Regional Forest Control Units under the State Environmental Inspection, which is subordinated to the Ministry of Environment, are responsible for implementation and extension of services, including compliance of rules and regulations under the legal act by public and private forest owners.

Another institution with a large influence on public-owned forests is the General Forest Enterprise. The institution is subordinated to the Ministry of Environment and is responsible for the management of forty-two commercial State Forest Enterprises. Forest Enterprise is the basic forest management unit responsible for the implementation of forest management plans in state forests. State Forest Enterprises are responsible for all forest activities related to regeneration, tending and protection of forests; forest utilization including harvesting operations, construction and rehabilitation of production facilities and buildings; road construction and maintenance of land drainage systems; recreation and equipment; and all other forest-related activities. Forests that have been set aside for privatization under the land reform are not being managed at all at the moment, and felling is forbidden in these areas.

Though little has been done to assess the most appropriate instruments to enhance woodfuel production in Lithuania, and to evaluate fiscal and economic implications of such instruments, the country's efforts to conform to policies and practices of the European Union could come as an advantage to bioenergy utilization. Yet, as further discussed in the last section, this is no guarantee and will depend on combined national efforts towards policy integration, energy efficiency and renewable energy, and the focus put on the promotion of bioenergy options.

7.3. FUELWOOD UTILIZATION IN LITHUANIA

Total biofuel consumption in Lithuania amounted to 3.3 million m^3 in 2001 (Department of Statistics to the Government of the Republic of Lithuania, Statistics Lithuania 2002) which generated some 7.3 TWh of energy (Ministry of Economy of the Republic of Lithuania, Energy Agency 2002). Biofuels were mostly used in the form of firewood, in addition to being used as sawdust briquettes, peat and other primary solid fuels. The large majority of the biofuels, about 90 per cent, were used for heating the households. The rest was used in commercial and public services, industries and other minor applications. Of the total, only a very small portion was used in district-heating plants (Energy balance 1999, Statistics Lithuania 2000). As mentioned before, the utilization of biomass for district heating is recent and started first in 1994.

Of the residues generated in sawmills, half goes to households, some 17 per cent are used in boiler houses, 17 per cent are used internally in the sawmills and 13 per cent goes to pulp and board industries. When it comes to the evaluation of total wood waste potential, the figures are less trustworthy. In any case, for strategic purposes, technical and economic availability for collection and transportation to boiler houses or densified woodfuel production plants needs to be considered together with the introduction of new practices and construction of new infrastructure.

It is important to notice how market forces are rapidly leading to new attitudes and practices. One interesting development is the increasing production of wood pellets and briquettes. Figure 7.1 illustrates this development in the last few years. From a very small production in the early 1990s, briquette and pellet production have expanded very rapidly. Most of these products are exported to Germany, Denmark, Sweden and Norway. In the local market, the price of pellets is about half of what it can reach in export markets even if prices have doubled since 1994. Thus raw materials that were simply being wasted before have now found a competitive market as a result of market forces and changing energy policies in Europe.

In principle, the expansion of wood-based industries tends to generate more residues, thus there should be no need for conflict among different users. However, competing uses for waste from processing industries, including producers of densified fuels, board factories and pulp producers, leave no optimism for large amounts of residues to be left for heating plants. Truly, the wood-processing industry in Lithuania will have to be restructured to be able to compete in open European markets, but this will also lead to a larger internal use of residues to dry the products. All in all, expansion of logging activities and general restructuring and efficiency improvement of sawmills shall result in an increase between 10 and 20 per cent only in the amount of residues generated from wood processing in the near future.

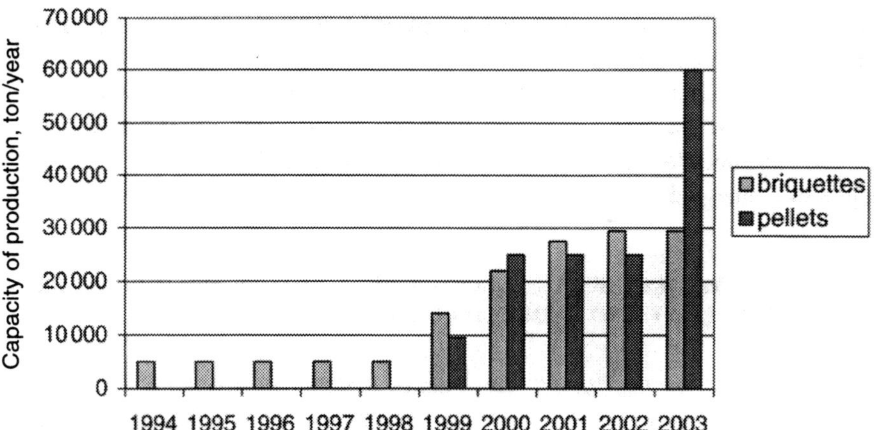

Figure 7.1. Pellets and briquettes production in Lithuania 1994–2003. Source: Andersson and Budrys (2002).

Thus, to achieve sustainability in the supply of wood fuels, especially in connection with conversion of oil- and coal-fired boilers in district-heating systems to biofuels, wood fuels supplied directly from forests will be very much needed. A significant shift towards greater utilization of wood-based energy will also require modifications in present policies including taxes and subsidies on energy sources. Careful attention needs to be paid to avoid potential conflict of interests between agricultural and forest activities, as well as wood-based industries in general.

7.4. DEMONSTRATION PROJECTS IN ROKISKIS FORESTS

Initial observations give the impression that the forest sector in Lithuania is resistant to change and thus not open to the new activities and routines needed to foster bioenergy use. A more attentive observation, however, indicate that the barriers to overcome are more complex than simply conservative thinking. Lack of incentives and coordinated policies, lack of established management practices for harvesting and marketing residues, weak internal demand, the small size of private forests as well as their risk aversion and lack of capital are some of the most difficult issues that need to be addressed.

As a starting point in this process, a project was designed to address the various elements of bioenergy systems in the Lithuanian context using the Swedish experiences as a reference track. A number of studies and assessments were carried out. The idea was to collect enough local information for the establishment of demonstration projects in Rokiskis public forests, which would then provide the basis for capacity building and dissemination of know-how related to bioenergy systems in the country. Some 15 to 20 per cent of the total national fellings are performed in this region, of which approximately half takes place in state-owned forests and the other half in private properties.

Table 7.1 summarizes the main studies that were carried out in the state-owned Rokiskis forests, indicating questions that these studies were striving to answer and the main methodologies used. The following section addresses some of the major results that emerged from the various studies.

7.5. NEW TECHNOLOGIES AND MANAGEMENT PRACTICES FOR HIGHER PRODUCTIVITY AND REDUCED COSTS

The motivation for developing fuel supply systems horizontally and vertically integrated with forest activities include opportunities for further development of the traditional activities themselves and future industrial expansion. Section 7.4 summarized some of the studies carried out with the purpose of evaluating resource

Table 7.1. Assessing forest fuel production in Rokiskis forest enterprise

Study	Main questions being asked	Methodologies used
Estimation of forest fuel resource potential	• How much residue are forest activities generating today? • What are the prospects in the coming decades? • How much residue can forests produce on a sustainable basis with improved methods and technologies, which can be used as fuel in energy generation?	• Inventory of mature stands in final cutting areas • Forecast of mature stands of various species in the coming two decades • Evaluation of nonmerchantable wood in final cuttings, commercial and precommercial thinnings
Technology assessment of forest fuel production	• How do different technologies and practices affect the potential extraction of forest fuel? • What needs to be changed in the felling operations as the intensity of cuttings increases, and how should that be done? • How does forest fuel extraction affect overall forestry operations? • What management practices can be used to improve the productivity of forests?	• Investigation of intensity of precommercial and commercial thinnings and final cutting of different species in sites of various types • Comparison of traditional Lithuanian thinning standards with Swedish precommercial thinning and biofuel outtake regulations • Evaluation of the integration of different technologies in the traditional context • Modeling stands and species composition to optimize productivity
Economic assessment of forest fuel production	• How much does the extraction of forest fuel cost in varying circumstances (different technologies and practices)? • What is the cost structure of the logging operations, product preparation and transport of fuel? • How sensitive is the cost to variations in distance to the boiler house? • How does forest fuel extraction affect the overall economy of forestry activities?	• Evaluation of workday and time expenditure of worker in various activities and applying different technologies, followed by monetary evaluation of work time • Evaluation of monetary expenditures for producing and thinning plot, transport from extraction to road, chipping and delivering to boiler house • Monetary evaluation of materials and tangibles • Assessment of profitability of forest fuel production and sensitivity analysis
Evaluation of ash handling methods	• How does wood ash influence mineral contents (e.g. heavy metal content) and microorganisms in forest soils? • What is the intensity of CO_2 depletion in the soils? • Should forest fuel ash be used as compensatory fertilisers in the forest soils and to what extent?	• Fertilization demonstration projects • Evaluation of soil's chemical and biological properties • Study of soil microflora and bioactivity • Verification of soil ground vegetation to evaluate species composition and diversity

Source: Based on *Studies on forestry, technology and economy of forest fuel production in Rokiskis forest enterprise* prepared by the Lithuanian Forest Research Institute.

potentials and models for forest management. The models are aimed at a sustainable exploration of residues for energy purposes while also improving traditional forestry activities. Analyses and tests were conducted to screen the most cost-efficient methods and most suitable equipment for local conditions, and to verify the best ways to introduce the new practices. This section provides a brief review of the results obtained.

The extraction and processing of logging residues, such as round wood or small trees, for fuel purposes can technically be carried out in many different ways. The amount of fuel supply depends largely on the size of the felling operations, level of mechanization and how forestry activities are carried out. The main factors influencing the cost of forest fuel production for boiler houses include worker skills, methods used for felling, intensity of extraction, extraction distance, actual productivity of the wood chipper and transportation distance to boiler houses. The extraction costs are lowest in final cuttings due to the concentration of fuels.

Precommercial thinnings are first of all aimed at improving conditions for tree growth, but may yield a profit if aimed at fuel production. This will depend on stand conditions, productivity of machinery and labor, and fuel prices. Considering the effects on the whole forest productivity, the option should be economically attractive even if the chip price only covers production costs. In fact, producing forest fuel may considerably reduce the costs of precommercial thinnings (see Table 7.4).

More time is used to handle the felled trees in precommercial stands than in commercial or final cuttings. This is due to the small diameter breast height (dbh) and large number of stems and longer distance to the technical corridors. In addition, the collection of trees is more complicated due to the remaining stands. Forest fuel production in precommercial thinnings using hand tools diminishes the losses by 61 Lt/ha as the income for the forest fuel covers some of the traditional cost for this measure. Motor chainsaw with felling handle tools should only be used in precommercial thinning stands where silvicultural measures have been carried out before or in conflict stands where dbh are no less than 5 cm.

Forest fuel production in precommercial thinnings is a novelty for foresters and workers in Lithuania. The forest workers can increase their productivity by 30 per cent when using machinery. Even when handle tools are used, such as in precommercial thinnings, and additional operations need to be performed, a productivity increase of 20 per cent can be achieved among workers performing tree cutting and storage operations. But there is no sufficient knowledge of working under these new technological requirements yet. Observations indicated that the psychological attitude of the worker towards the new technologies and practices affected the results significantly but, in general, comfort levels could be achieved after a couple of days of experience, which indicates that productivity increases can be made normative rather rapidly.

The cost of forest fuel extraction depends on many factors. In Rokiskis forests, the structure of forest fuel cost was analyzed following logging operations and various cost factors from cutting to handling and transporting to boiler houses. First of all, the cost will depend on the type of cuttings. The cost is lowest in final cuttings, and highest in precommercial cuttings for reasons explained earlier. The total cost of forest fuel produced in commercial thinning with integrated methods is 34–35 Lt/solid m^3 (excluding overhead costs) chipped and transported to the heating plant, while in the clear cutting the cost is 30 Lt/solid m^3.

Another important cost factor is machinery productivity. When the productivity of the chipper is increased from 10 to 50 m^3/h, the fuelwood cost may fall approximately 1.5 times. A third cost factor is extraction distance. With an increased transportation distance from 5 to 30 km, the cost increases by more than 50 per cent. Forest fuel transportation cost is greatly influenced by the transport equipment used. When transporting by MAZ 5516 with the trailer of 45 m^3 for 30 km, forest fuel costs decrease by 25 per cent in comparison with transportation by the tractor T-150 with a trailer of capacity 25.3 m^3.

In short, chipping accounts for 38.7 per cent of the final forest fuel cost in final cuttings, this being the most significant cost factor. Transportation and extraction make up about 23–24 per cent of the total cost each (see Figure 7.2). A closer inspection indicates that machinery costs excluding fuel account for 60 per cent of the cost composition while raw material costs amount to 17.6 per cent of the total only (see Figure 7.3). This means that increased productivity of machinery through longer hours and more days of operation per year can have a substantial impact on the final fuelwood cost.

Lithuanian stand thinning models already have high productivity. There is no particular need for radical changes here but some modifications of thinning systems

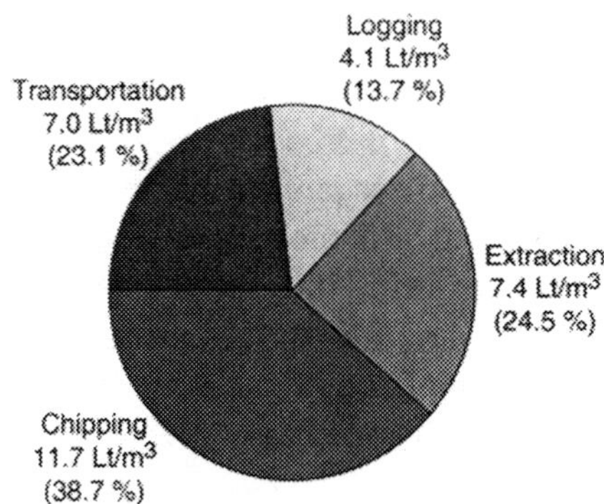

Figure 7.2. Fuelwood production costs according to operation factors. Source: Andersson and Budrys (2002).

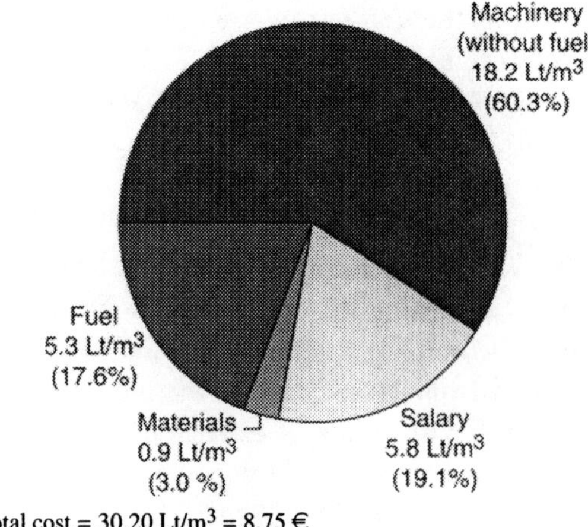

Figure 7.3. Fuelwood production costs according to inputs. Source: Andersson and Budrys (2002).

can bring advantages. This refers not to the intensity of cutting, but to the aspects of stand species composition. In addition, it is not advisable to extract all nonmerchantable wood from the forests in Lithuania, not even in commercial thinnings. Forest soils can be exhausted after taking out the wood without compensation for fertilization. Some experts in Lithuania are in favor of extracting all merchantable branches and up to 30 per cent of nonmerchantable branches. However, it is advisable that the amount of nonmerchantable branches used as fuelwood in thinnings be reduced down to 20 per cent. At this moment, there are no obstacles for forest fuel extraction in forestry legislation.

A number of advantages were observed in cuttings with integrated forest fuel production. For example, in final cuttings, lesser quantity of branches were used for technical corridors and this avoided the so-called "widening" of technical corridors. In conflict, for stands of spruce with broadleaves, some amount of broadleaves can be left to reach 15 years of age to accumulate additional amount of wood and keep the productivity of the stand at a homogeneous level. In general, the amount of produced forest fuel increases with integrated forest fuel handling technology. Table 7.2 shows the observed variations in the amount of cubic meters of forest fuel obtained per hectare when applying traditional and integrated technologies in commercial and final cutting. Integrated technologies can help increase the amount of fuel extracted to a factor of 15.

The smallest amount of forest fuel was produced in typical commercial thinnings where technical corridors had been made in the previous cutting. The amount of forest fuel produced in clear cutting areas was 58–112 solid cubic meters per hectare.

Table 7.2. Produced forest fuel using traditional and integrated technologies in commercial and final cutting (in m^3 solid volume/hectare)

Cutting category	Traditional technology				Integrated technology				Forest fuel produced increased with factor
	Industrial wood		Fire wood	Total volume	Industrial wood		Forest fuel	Total volume	
	Long	Short			Long	Short			
Commercial thinning	2	23	0.5	25.5	2	23	5	30	10
Commercial thinning	19	57	1.4	77.4	19	57	21	97	15
Commercial thinning	30	32	6	68	30	32	31	93	5
Sanitary cutting	32	48	6	86	32	48	12	92	2
Clear-cutting	75	160	17	252	75	160	58	293	3
Clear-cutting	114	79	9	202	114	79	61	254	7
Clear-cutting	80	122	25	227	80	122	112	314	4

Source: Andersson and Budrys (2002)

The largest amount of forest fuel was produced in aspen stands with dense understorey while smaller amounts were produced in spruce and pine stands. The amount of forest fuels increases with extraction from precommercial thinning stands, the average figure of extracted forest fuel from the precommercial thinning stands being 56 m^3 solid volume per hectare.

The production cost for traditional industrial assortments decreases with integrated forest fuel handling. Due to more rational handling, some cost for handling the traditional industrial assortment turns over to the forest fuel assortment at the same time as the total productivity increases. Table 7.3 shows the gains obtained at each step. As shown, using integrated technology, the costs for extraction of industrial wood can be reduced by up to 15 per cent while fuelwood is also generated.

Comparing with traditional technology, when all branches and tops are used for technical corridors, the technology with integrated forest fuel production is very promising. However, it requires a new way of thinking about silviculture. The integration of forest fuel handling requires a mobile drum wood chipper for acceptable decomposition of the forest fuel. With a better knowledge of the methodology and more skilled personnel performing the operations, the results from the integrated forest fuel handling methods can be further improved.

7.6. CONTINUING EFFORTS IN THE BALTIC SEA REGION

Energy questions are receiving significant amount of attention in the Baltic States, particularly energy conservation and the use of renewables aimed at increasing

Table 7.3. Direct costs for wood produced and extracted to roadside using traditional and integrated technology in commercial and final cutting (in Lt/solid m^3)

	Traditional technology	Integrated technology		
Cutting category	Industrial wood incl. fire wood (Lt/solid m^3)	Industrial wood (Lt/solid m^3)	Forest fuel (Lt/solid m^3)	Reduced cost for industrial wood, (%/m^3)
Commercial thinning	27	23	15	15
Commercial thinning	28	24	15	14
Commercial thinning	29	27	14	7
Sanitary cutting	26	25	14	4
Clear-cutting	18	17	11	6
Clear-cutting	19	17	11	11
Clear-cutting	22	20	10	9

Note: 1€ = 3.45 Lt
Source: Andersson and Budrys (2002)

Table 7.4. Total economic efficiency using integrated and traditional technologies

	With integrated forest fuel production			Traditional technology		Improved efficiency in production	
Cutting category	Income	Cost	Profit	Income	Cost	Profit	Lt/ha
Pre-commercial thinning (R)	2184	2442	−258	–	319	−319	61
Pre-commercial thinning (M)	2184	2860	−676	–	476	−476	−200
Commercial thinning	1972	766	1206	1790	699	1091	115
Commercial thinning	6907	2531	4376	6072	2382	3690	686
Commercial thinning	7129	2681	4448	6074	1961	4113	335
Sanitary cutting	7664	2412	5252	7310	2243	5067	185
Clear-cutting	22 287	5639	16 648	20 399	4639	15 760	888
Clear-cutting	22 162	5200	16 962	19 992	3882	16 110	852
Clear-cutting	22 640	7426	15 214	18 810	4978	13 832	1382

Note: Both alternatives include handling of the industrial assortment; the forest fuel is extracted, chipped at roadside and transported 15 km to a heating plant.
R = using manual tools;
M = using motor chain saw with felling handle.
1€ = 3.45 Lt
Source: Andersson and Budrys (2002)

security of supply. In fact, energy intensity in Lithuania is three times higher than the EU average and a potential for more than 30 per cent increased efficiency has been identified. Most of this potential improvement exists in the household sector. (Klevas and Antinucci, 2004).

In Lithuania, most sources of financing for renewables are focused on tax reductions and guarantees for energy saving investments (BASREC and Nordic Council of Ministers, 2002). As candidate member to the EU, Lithuania was entitled to receiving financial support from Structural Funds, together with Latvia and Estonia. The Lithuanian Energy Institute participated in a project to promote energy efficiency and renewables in the preaccession phase (Klevas and Antinucci, 2004). Lithuania is now a full member of the EU and has adopted the EU directive on renewables. Lithuania's targets include 7 per cent renewables by 2010 and the decommission of its nuclear power plant. Regional programs, for example to develop bioenergy options may provide an important complement to top down energy strategic plans, while also attracting other financial sources through bilateral or multilateral cooperation.

More specifically, the Baltic Sea Region Energy Co-operation, BASREC, created within the EU in 1999, addresses energy issues in the Baltic. The Nordic Council of Ministers and BASREC developed the Bio2002Energy project with the purpose to gather overall information of bioenergy prospects in the Baltic Sea Region (BASREC and Nordic Council of Ministers, 2002). Some of the major overall conclusions include:

- Bioenergy still need financial support to be viable due to high investment requirements;
- Joint implementation, in the modes defined under the Kyoto Protocol, will serve as a major drive for bioenergy projects in the region in the coming years;
- Harmonization of energy and environmental taxes should be pursued to provide good competitive ground for bioenergy-related companies;
- The development of energy crops should be promoted based on regional know-how and experience and in face of new options to the agricultural sector;
- Coordination of actors in various sectors such as environment, agriculture, forestry and energy is needed to minimize costs of fuel procurement, utilize resources more efficiently and guarantee the sustainability of ecosystems.

When it comes to promoting renewable energy technologies in the EU, member states have applied different strategies to reach their targets and, though some instruments have proved more successful than others, none of them can be said to be particularly superior (Reiche and Bechberger, 2003). Thus integration into the EU does not bring automatic answers when it comes to renewable energy strategies, bioenergy included. There are plenty of experiences in neighboring countries to serve as starting points, but there is still need for a national policy focused on the particular potential and conditions of the country.

The project described here has helped to indicate the existing potential for biofuel production in Lithuanian forests, and to point management practices that can contribute to harvesting biofuels efficiently. There is no doubt that the country can count on a large biomass supply, which offers a great resource base for bioenergy options. Targeted incentives and a strong institutional framework are now needed to guarantee a stable development of the demand side. Only then can the country's potential be fully realized and bioenergy systems mature on a sustainable basis.

The increasing utilization of biomass in district-heating systems has contributed to creating a continuous and more concentrated demand for biofuels while also helping demonstrate applications and various benefits of bioenergy. However, we should keep in mind that, even at the European level, the future of district heating systems and their role in liberalized markets vis a vis gas networks, for example, are not regulated, and none of the EU directives consider district heating systems in particular (Grohnheit and Mortensen, 2003).

One strong motivation for support of bioenergy from the political side is obviously the fact that the use of biofuels creates employment for farmers, forest workers and entrepreneurs in plant operations. In fact, the importance of these activities for rural development and living conditions in the countryside should not be underestimated. Employment is also created in the equipment manufacturing industry. For a country that is strongly dependent on energy imports such as Lithuania, bioenergy offers a great opportunity to improve energy supply security based on national resources. In addition, the harvesting of forest fuels can be a driving force for improved forest management, and thus enhancement of the economy of forest industries. For example, biofuel demand will favor various industrial assortments to the extent that many stands that are normally not managed today will be so thanks to the possibility of extracting forest fuels.

Energy development authorities at national and regional levels in Lithuania still lack the necessary information and knowledge to outline a strategy for woodfuel procurement. The methods applied at municipal level to evaluate the energy potential derived from forests vary significantly and can sometimes be contradictory. Cooperation and coordination of efforts between forest and energy sectors are necessary to ensure the development of a strategy and the integration of forest fuel handling into common forestry practices.

REFERENCES

Andersson, L. & Budrys, R. (2002) Integration of forest fuel handling in the ordinary forestry, Forest Department in the Ministry of Environment of the Republic of Lithuania, Swedish Energy Agency, National board of Forestry, Vilnius.

Baltrusaitis, A. & Andersson, L. (2000) Potential biofuel use in Lithuania, Department of Forest and Protected Areas in the Ministry of Environment of the Republic of Lithuania, Swedish Energy Agency, National board of Forestry, Vilnius.

BASREC and Nordic Council of Ministers (2002) Bio2002Energy – Development of the Use of Bioenergy in the Baltic Sea Region, Report.

Grohnheit, P.E. & Mortensen, B.O.G. (2003) Competition in the market for space heating. District heating as the infrastructure for competition among fuels and technologies in *Energy Policy* **Vol. 31**(9), Elsevier, pp 817–826.

Klevas, V. & Antinucci, M. (2004) Integration of national and regional energy development programs in Baltic States in *Energy Policy* **Vol. 32**(3), Elsevier, pp 345–355.

Lithuanian Forest Research Institute (2001) Assessment of Forest Fuel Production Technological and Economical Aspects in Pre-Commercial Thinnings, Report.

Lithuanian statistical yearbook of forestry (2000) Centre of Forest Economics.

Reiche, D. & Bechberger, M. (2004) Policies differences in the promotion of renewable energy in EU member states in *Energy Policy* **Vol. 32**(7), Elsevier, pp 843–849.

Part III

Promoting Bioenergy Utilization

Chapter 8

Potential for Small-scale Bio-fueled District Heating and CHPs in Sweden

Thomas Sandberg and Knut Bernotat

8.1. AIMING AT SUSTAINABLE ENERGY SYSTEMS

The share of renewable energy is high in the Swedish energy system. In 2001, 177 TWh or 44 per cent of the total energy consumption in the country came from renewable sources. In the residential, service and industrial sectors, this share was as high as 58 per cent. Since 1991, bioenergy is the largest renewable source in Sweden and, in 2001, 98 TWh were generated from this source. This compares with 79 TWh from hydro and 0.5 TWh from wind (Swedish Energy Agency, 2002a; see also Ling, Chapter 3).

Even if these are impressive figures, some 122 TWh remain to be converted to renewable sources, not to mention the other 92 TWh in the transport sector. Scenarios developed for the year 2050 indicate that this is a manageable task for Sweden (Elforsk, 1996; SAME, 1999). The crucial point is to speed up the process and to make it cost efficient. Although the resources and technologies are there, administrative, economic and political structures delay the shift towards a sustainable energy system.

District heating supplies 46 TWh of the 93 TWh heat consumption in Sweden today. Nearly all this energy emanates from large-scale units connected to large-scale district-heating systems. Some 32 TWh are generated from biomass. Due to the large availability of electricity from hydro and nuclear sources, only 5 TWh of electricity is being produced in combination with heat. We argue that district heating, supplied by combined heat and power (CHP) production based on bioenergy, holds an important key to the shift towards larger use of renewables in Sweden in the next two decades. Most probably, this is also true for many other countries with moderate climate, where there will always be a large need for heating.

There are many ways to enhance the utilization of bioenergy in Sweden. In general, this includes:

- Continuing the conversion from fossil fuels to biofuels;
- Increasing the power production in sites where CHPs are installed and fueled by biomass;

Bioenergy – Realizing the Potential

- Introducing power production in sites where only heat is produced from biomass;
- Enlarging existing district-heating grids; and
- Building new district-heating grids and CHP units fueled by biomass.

Since large cities and many towns in Sweden already have a district-heating system, the remaining potential is in smaller places with a few thousand down to even less than one thousand inhabitants. Therefore, in this chapter, we focus on small-scale district heating grids supplied by small-scale CHP fueled by biomass. Our task is to calculate the potential for such systems in Sweden.

We first present a method to estimate the potential for small-scale district heating. The essence of the method is to identify clusters of buildings where the heat demand is large enough to justify small grids. We use this method to evaluate the potential of a small region in southeast Sweden, and of three counties in different parts of the country. Finally, we extrapolate the results to make a rough estimation of the potential for new district heating systems in the country as a whole, and to evaluate the potential generation of electricity from the CHP units.

8.2. A METHOD TO ESTIMATE THE HEAT DEMAND

We start from the demand side, evaluating the demand for heating services. The basic notion is to identify geographic areas (clusters) where the heat demand is large enough to justify installing a local small-size heat and power production. But how to identify those clusters with different residential, commercial, industrial and public buildings, and how to estimate their heat demand?

When looking for the potential demand, we need to differentiate between theoretical and practical potential. The theoretical potential is the overall estimate of the demand for heat in a given area. It is based on the estimates for each individual building in the area. The theoretical potential includes all buildings and thus even those where it is not currently technically or economically feasible to use small-scale district heating. The theoretical potential is an important reference because technical and economic conditions change over time and move the border for feasible applications. The theoretical potential also gives the individual, nonaggregated data for estimating the practical potential (Bernotat, 2002).

The practical potential in an area is defined as the total heat demand of buildings, which are technically and economically feasible to connect to a small-scale district heating and CHP. The practical potential is found by merging the theoretical potential of single buildings to larger units until a suitable size is found. Figure 8.1 illustrates the relation between the theoretical and practical potential over time.

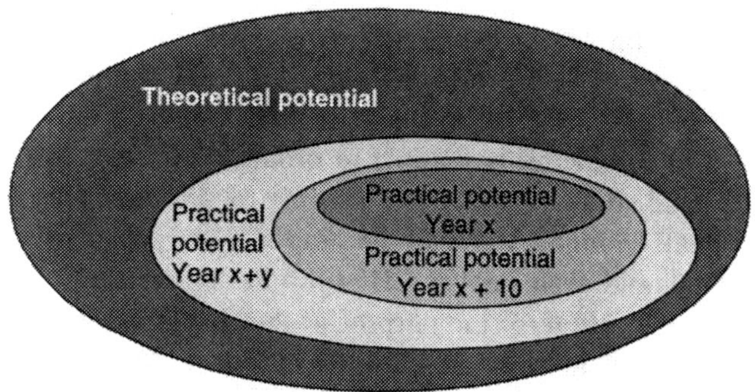

Figure 8.1. Theoretical potential and practical potential over time. Source: Bernotat, 2002.

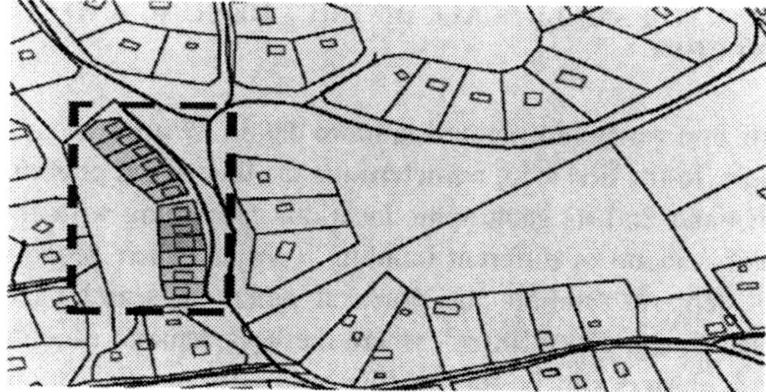

Figure 8.2. Finding the practical heat demand potential in a residential area. Source: Bernotat, 2002.

Figure 8.2 shows how this translates into the reality of suitable clusters when they are visualized with the support of the GIS tools applied in this study.

The actual estimation of the potential is done in two steps. First, the heat consumption of every single building is estimated using data from the land survey register on building type, year of construction, floor space and the precise location. The survey data on the average heat demand per square meter for buildings of different types, age and location are then added. Second, the estimated heat demand of the individual buildings in the focused geographic cluster is aggregated with the help of Geographical Information System (GIS) tools. The shape and size of the clusters within an area can be varied, as also the minimum heat demand required for the clusters in order to match preferences and predefined economies of scale[1].

[1] The method used in this study is more thoroughly described in Bernotat (2002), Bernotat and Sandberg (2002), and especially in Sandberg and Bernotat (2003).

One crucial question that remains always is what minimum heat demand is required to justify the infrastructure investment. In the clusters with more than 2 GWh heat demand, we are close to the point where, already today, it is not only technically but also commercially feasible to produce heat and power. At around 1 GWh heat demand, it is commercially feasible to produce heat if the district-heating grid is small enough. A crucial question is when power production equipment for such small and even smaller CHPs will be commercially available. These figures serve as an indication for the purpose of this analysis but need to be further investigated in each case, not least in relation to the grid size and the availability of biomass to fuel a CHP, as we have in mind.

8.3. POTENTIAL FOR SMALL-SCALE DISTRICT HEATING AND CHP IN A SMALL REGION

To demonstrate and verify our method in more detail, we apply it to a small region well known to us. In the first step, we determine the theoretical potential, that is, the overall heat demand and its geographic location, continuing with the distribution of the total heat demand of different building types and their geographic location. In the second step, we estimate the practical potential, searching out the small geographic clusters, e.g. $500 \times 500\,m^2$, where the heat demand exceeds for example 0.5 GWh. Those clusters can be the starting points for small-scale district heating and CHPs.

The $36 \times 48\,km^2$ region is located in the middle of the county of Kalmar, southeast Sweden. It is a rural district with 8000 inhabitants, thus only 5 inhabitants per km^2. There are 2000 people living in the largest urban agglomeration, and seven villages with populations between 300 and 600 inhabitants. This small region is mostly forested with an abundant supply of wood waste. In addition, there is a small agricultural district around the largest center and some of the other villages, which can provide biomass in the form of agricultural residues. There are ten hydropower stations in the two main rivers of the region, producing approximately 25 GWh annually. Solar and wind energy is very marginal. There is one new, small district-heating system of 0.8 GWh using wood pellets.

The theoretical heat demand potential found for this region is 102 GWh, based on estimates for every single building, and implying an average heat demand per inhabitant near the national average 10 MWh. Figure 8.3 shows the geographic distribution of the theoretical potential in the region, where relevant concentration is found in less than ten places. Obviously, there is a strong correlation between the distribution of population and activities performed in the region and what is observed in this figure.

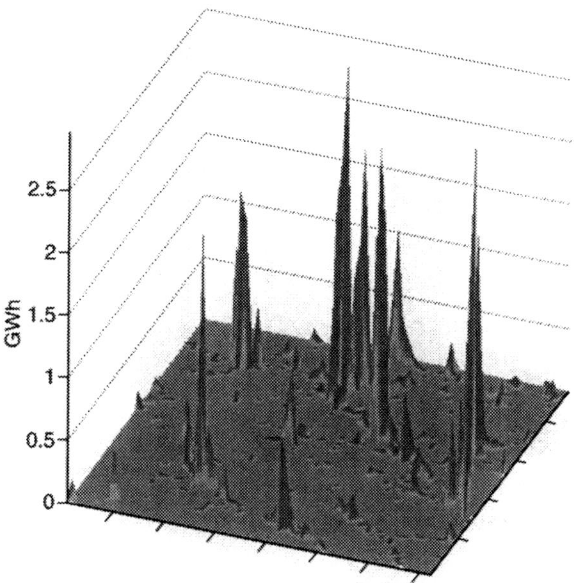

Figure 8.3. Location and size of estimated heat consumption in all building types in 500 × 500 m^2 clusters in a small region of Kalmar.

Of the total 102 GWh, 77 GWh is the demand of one- and two-family houses, 7 GWh from multidwellings, 13 GWh from industrial, 2 GWh from commercial and 3 GWh from public buildings. This distribution was anticipated as small houses dominate in rural districts. Thus 75 per cent of the theoretical potential comes from single dwellings. Multidwellings are found in only eight places with at least 300 inhabitants, and industrial and public buildings are more scattered. As much as 25 per cent of the heat demand is found in the larger buildings, which facilitates the search for starting points for the district-heating grids.

Stepping from the theoretical to the practical potential, we focus our interest on small geographic clusters where the heat demand exceeds a certain limit. We have used a quadratic shape, the four cluster sizes 250 × 250 m^2, 500 × 500 m^2, 750 × 750 m^2 and 1000 × 1000 m^2, and a minimum heat demand of 0.5 GWh. Table 8.1 shows the number of clusters found for each range of heat demand. Obviously, when the cluster area increases, more clusters reach the minimum stipulated limit of 0.5 GWh. At the same time, adjacent clusters merge.

Focusing on single dwellings, the share of the heat consumption included in the clusters in relation to the total 77 GWh increases as the clusters are enlarged, going from 16 per cent for the 250 × 250 m^2 clusters to 65 per cent for the 1000 × 1000 m^2 clusters. Consequently, the average heat consumption per cluster increases from 0.6 GWh in the smallest cluster size to 1.6 for the largest cluster size. At the same

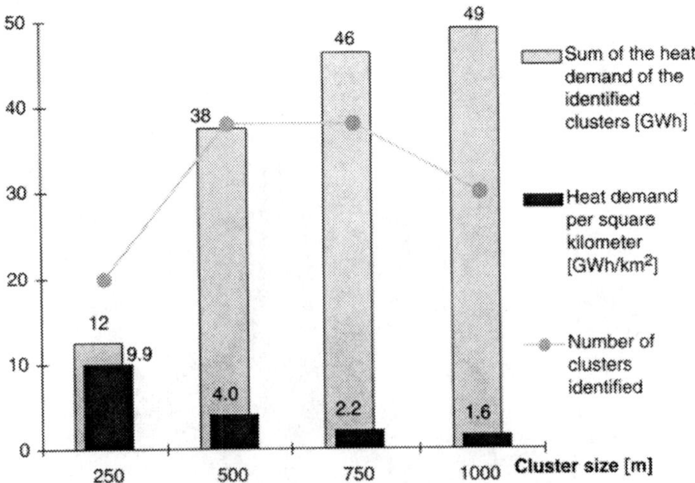

Figure 8.4. Heat demand characteristics in relation to the cluster size for single houses.

time, the average heat consumption per km^2 in a cluster decreases from 9.9 GWh per km^2 in the 250×250 m^2 clusters to 1.6 in the 1000×1000 m^2 clusters (see also Figure 8.4).

Before leaving our small region, we will also summarize the heat demand for all building types. We limit the analysis to 500×500 m^2 clusters and a minimum heat demand of 0.5 GWh per cluster. In this case, we arrive at 53 clusters, which together account for 64 of the total 102 GWh heat demand in this small region. Table 8.1 indicates the number of clusters found for each range of heat demand and allows an easy comparison with the case where only single houses are considered. The inclusion of all building types results in a larger number of clusters with enough heat demand to justify combined heat and power already today. Moreover, the nonresidential buildings with their larger average heat demand can serve as crystallization points for the heat grid. Many of them also have production facilities that can be of some

Table 8.1. Number of clusters found in each range of heat demand

	Range		
Cluster size/single houses	0.5–1.0 GWh	1.0–2.0 GWh	Larger than 2.0 GWh
250×250 m^2	20	–	–
500×500 m^2	25	11	2
750×750 m^2	22	10	6
1000×1000 m^2	13	8	9
All buildings 500×500 m^2	24	23	6

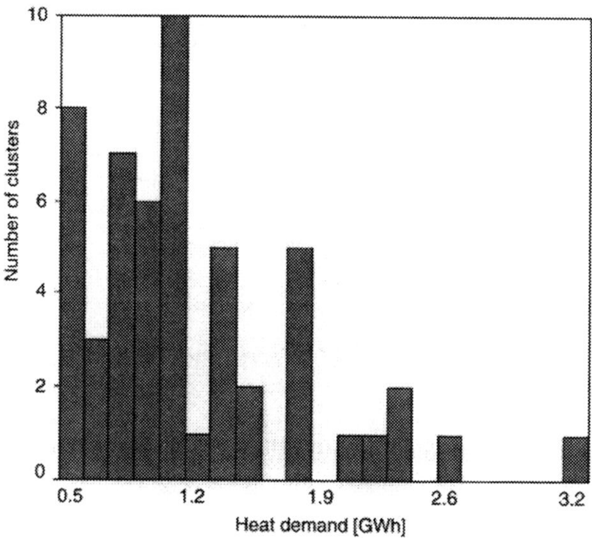

Figure 8.5. Distribution of the 500 × 500 m² clusters in relation to heat demand – all building types and demand over 0.5 GWh annually.

use. Figure 8.5 shows the distribution of the clusters found, and their heat demand in more detail.

A good half of the heat demand from single houses in the area, and some 85 per cent of all heat demand from other buildings are captured in the clusters found. Continued technical development and increased taxation of external costs of nonsustainable energy sources can make more clusters rapidly ready for CHP, while also attracting investors to build the necessary infrastructure.

8.4. POTENTIAL FOR SMALL-SCALE DISTRICT HEATING IN THE COUNTIES OF KALMAR, ÖREBRO AND VÄSTERNORRLAND

We now apply our method on three counties in Sweden: Kalmar in the south, Örebro somewhat in the middle and Västernorrland in the north. Table 8.2 summarizes some general facts about the counties in question. When choosing counties for this analysis, we considered the degree of urbanization of the regions due to its positive covariance with district heating.

After estimating the total heat demand for each county, we restrict ourselves to the identification of the 500 × 500 m² clusters where the total heat demand for all building types is at least 0.5 GWh. To arrive at the potential for small-scale district heating and CHP, we subtract the existing district heating. The geographic location of the 500 × 500 m² clusters with at least 0.5 GWh is shown in Figure 8.6.

Table 8.2. Facts about the counties of Kalmar, Örebro and Västernorrland

	Kalmar county	Örebro county	Västernorrland county
Land area	11 171 km^2	8519 km^2	21 678 km^2
% of Sweden	2.70%	2.10%	5.30%
Population	236.501	273.822	249.299
% of Sweden	2.70%	3.10%	2.80%
Urbanization	51%	68%	50%
Population per square km (Sweden 21.6)	21.2	32.1	11.5

Source: Swedish Energy Agency, 2001.
Note: Urbanization is estimated as the share of the population living in the biggest town in each municipality.

The concentration of the population and thus of the heat demand in larger urban areas is obvious, and the very sparsely populated areas of western Västernorrland, and northern parts of Örebro are apparent.

Table 8.3 summarizes the results obtained for different ranges of heat demand in the three counties analyzed. Though the number of clusters seems impressive at first, they actually include areas that are already served by district-heating grids. These areas need to be now subtracted from the total. Once this is done, we estimate the small-scale district heating potential at 0.9 TWh in Kalmar, 0.9 TWh in Örebro and 1.4 TWh in Västernorrland. These figures indicate that a very significant additional portion of the heat demand in these counties can be met with small-scale district heating. These increments are equivalent to 42 per cent of the total heat demand in Kalmar, 28 per cent in Örebro and 47 per cent in Västernorrland.

8.5. THE POTENTIAL FOR SMALL-SCALE DISTRICT HEATING AND CHP IN SWEDEN

In the whole of Sweden, district heating supplies 46 TWh, or 49 per cent of the total heat consumption in the country. Though this is a very significant accomplishment, there is still a large untapped potential for small-scale district heating in the country as a whole. Our analysis of three counties illustrates that assertion. Considering clusters of $500 \times 500\,\text{m}^2$ and a minimum heat demand of 0.5 GWh, the potential is as high as 28 per cent in Örebro, 42 per cent in Kalmar and 47 per cent in Västernorrland compared with the total heat demand in these counties. Moving the limit to 1.0 GWh gives a potential 31, 22 and 39 per cent in the three counties respectively.

The numbers obtained are very impressive. For the purpose of illustration, if we use the percentage found in Örebro (28%) to estimate the total potential for

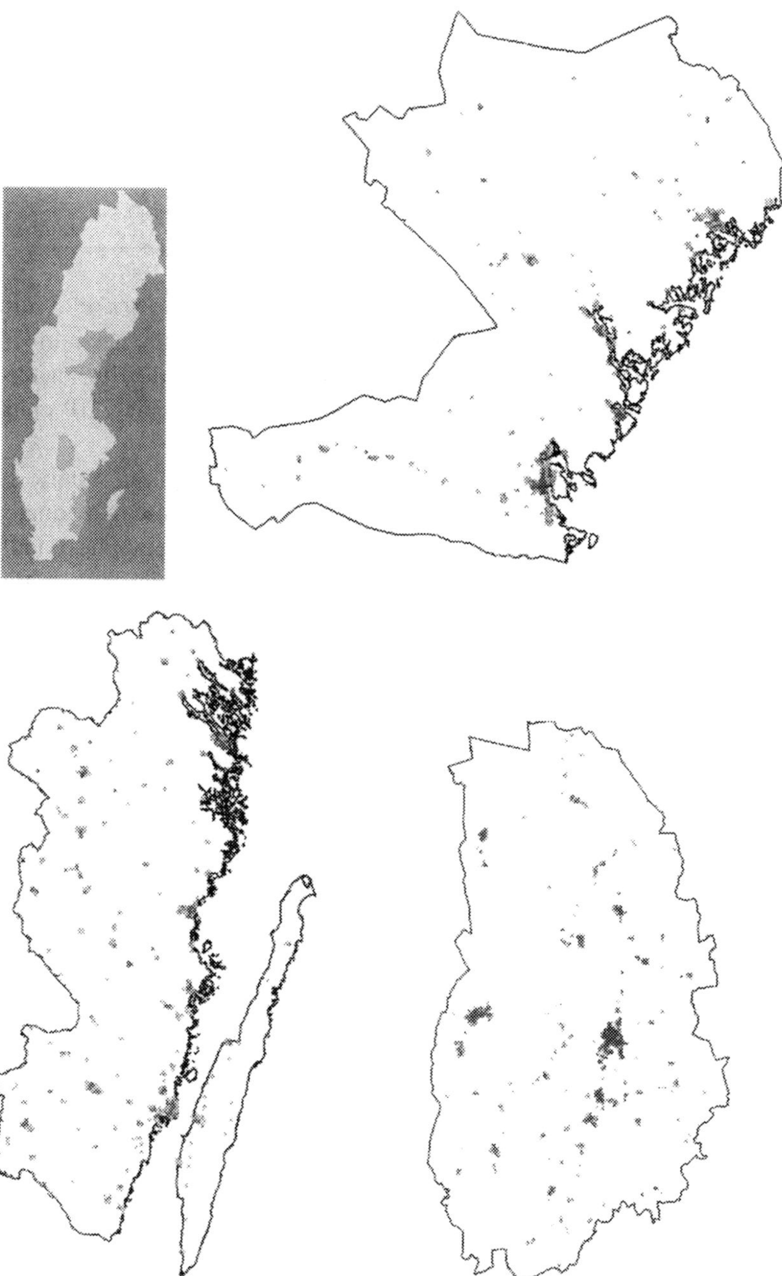

Figure 8.6. Location of 500×500 m^2 clusters with heat demand over 0.5 GWh in Kalmar (down left), Örebro (down right) and Västernorrland (upper right). (The location of the counties in Sweden is seen in the upper left.)

Table 8.3. Number of clusters found in each range of heat demand

County	Range		
	0.5–1.0 GWh	1.0–2.0 GWh	Larger than 2.0 GWh
Kalmar	300	257	266
Örebro	216	213	416
Västernorrland	310	272	349

additional small-scale biofueled district heating in Sweden, we arrive at additional 26 TWh for the country as a whole. Although the method applied here to verify the existing potential for new district-heating systems is basically fuel neutral, our starting point is that the heat should be supplied from small-scale CHP units fueled with biomass and connected to the existing electricity grid.

In areas without district heating or electric heating, the heat supply is decentralized today with a furnace in nearly each building. Once these individual systems are centralized into small-scale clusters, it is possible to cogenerate heat and power. In comparison with the dominating power supply today, the new contribution from each cluster and small CHP will be small, but a program to tap this potential can result in significant additions of heat and power to the system as a whole. The national power system will benefit from the new heat and power supply in two ways: through a decrease in the consumption of electricity for heating purposes, and an increase in the capacity for electricity generation from the new small-scale CHP units.

To calculate the consequences for the power system, one must contrast the added heat and power from biomass-fueled CHPs with the energy carriers that are being used today. We have two rough estimates, one for our small region in Kalmar and one for the whole of Sweden. In both cases we use $500 \times 500\,m^2$ clusters with at least 0.5 GWh heat demand.

In our small region, small-scale district heating and CHPs can substitute 64 GWh of today's heat consumption of 102 GWh. In one of the places in the region, heat is supplied by 60 per cent oil, 25 per cent electricity and 15 per cent firewood (Sandberg, 2001a, b). Applying those figures on the 64 GWh in the 53 clusters identified in the small region gives a substitution effect of 38 GWh from oil, 16 GWh from electricity and 10 GWh from firewood. The 16 GWh electricity are equivalent to a production capacity of 3 MW.

With an average heat demand of 1.2 GWh in each cluster and 3000 h operation time per year, we need a furnace with 400 kW heat capacity (Fredriksen and Werner, 1993). The power capacity can be assumed to be 100 kW or 5 MW for all clusters. With an operation time of 4000 h a year, the power production will be 20 GWh per year. Thus in our small region only, small-scale district heating fueled by biomass-based CHP can add 8 MW of power capacity, or 36 GWh of electricity to the grid.

For the whole of Sweden we have estimated the potential for small-scale district heating and CHP at 26 TWh. In 2000, the heat consumption outside district heating systems was 22 TWh or 42 per cent from electricity, 20 TWh or 39 per cent from oil and 10 TWh or 19 per cent from biomass, mostly firewood (SCB, 2001). Thus the consumption of electricity for heating can be reduced by 11 TWh (42 per cent of 26 TWh), which is equivalent to 2.2 GW of generation capacity.

At the same time that electricity demand for heating is reduced, the installed capacity and generation of electricity can increase. To substitute 26 TWh with heat from a small-scale CHP, an installed heat capacity of 8.50 GW is necessary. The power capacity can be assumed to be 2.15 GW which, with an operation time of 4000 h a year, gives 8.5 TWh (Fredriksen and Werner, 1993). Thus a broad program to install small-scale district heating based on bio-fueled district heating and CHPs could result in 19.5 TWh of added electricity out of an increased capacity of 4.35 GW. In addition, we should not forget the reduced distribution losses: for 19.5 TWh, losses are estimated at 1.5 TWh.

8.6. THE BENEFITS

Small-scale biofueled district heating and CHPs can make a substantial contribution to building a sustainable energy system based on renewable resources, as demonstrated in this study. In Sweden, a total of 21 TWh electricity can be added if the existing potential is tapped, which corresponds to 14 per cent more electricity in the grid, or an added installed capacity of 15 per cent. Electricity that is released from heating systems can be used more efficiently to provide other energy services. This is definitely a very significant development of the Swedish power system. Moreover 10 TWh oil and 5 TWh firewood are substituted by sustainable biomass. In countries with large heat demand and where district heating systems are not being as largely used, the contribution of such small-scale systems could be even greater.

Small-scale CHPs are interesting solutions in combination with other small-scale energy techniques. Throughout the year, solar energy and bioenergy complement each other in an ideal way due to seasonal variations of resource availability and energy needs. In local energy systems with small-scale hydro or wind power, bio-fueled CHPs can be a cornerstone.

The distributed power production has also the advantage to reduce the vulnerability of the power system. In the case of failure of large centralized units or due to distribution disruptions, the local power can be used to keep vital functions in operation. This will reduce the need for other back-up systems. In the cases where fossil fuels are being used for heating, the utilization of biomass-based systems will imply significant reduction of greenhouse gases. Finally, the establishment and operation of small-scale district heating and CHP units can help promote regional development, contributing to job generation for systems operations, for example, along the whole biofuel production chain, and related activities.

REFERENCES

Bernotat, K. (2002) Local production potential for small-scale heat and power in Sweden: the case of three regions, Department of Industrial Economics and Management, Royal Institute of Technology, Stockholm.

Bernotat, K. & Sandberg, T. (2004) Biomass fired small-scale CHP in Sweden and the Baltic states: A case study on the potential of clustered dwellings in *Biomass & Bioenergy*, **Vol. 27**(6), Elsevier, pp 521–530.

Carlsson, A. (2002) Considering external costs, Division of Energy Systems, Department of Mechanical Engineering, Linköping University.

Elforsk (1996) Ett uthålligt elsystem för Sverige (A Sustainable Electricity System for Sweden), Elforsk, Stockholm.

Frederiksen, S. & Werner, S. (1993) Fjärrvärme (District Heating), Studentlitteratur, Lund.

SAME (1999) Hållbar energiframtid? (Sustainable Energy Future?), Swedish Environmental Protection Agency, Stockholm.

Sandberg, T. (2001a) Biobränslebaserad kraftvärme i Alsterbro? (Bio-CHP in Alsterbro?), Department of Industrial Economics and Management, Royal Institute of Technology, Stockholm.

Sandberg, T. (2001b) Effektivare och miljöanpassad vattenkraft i Alsterbro (More Efficient and Sustainable Hydro Power in Alsterbro), Department of Industrial Economics and Management, Royal Institute of Technology, Stockholm.

Sandberg, T. & Bernotat, K. (2002) DG by CHP as a big step towards sustainability, *Proceedings from the 2nd International Symposium on Distributed Generation: Power Systems and Market Aspects*, Eds. Ackermann, T. & Knyazkin, V., Department of Electric Power Systems, Royal Institute of Technology, Stockholm.

Sandberg, T. & Bernotat, K. (2003) En metod för att uppskatta den svenska potentialen för småskalig biobränslebaserad fjärrvärme och kraftvärme (A Method to Estimate the Swedish Potential for Small-scale Biofuelled DH and CHP), Department of Industrial Economics and Management, Royal Institute of Technology, Stockholm.

SCB (2001) Energistatistik för småhus, flerbostadshus och lokaler (Energy Statistics for Dwellings and Non-residential Premises), SOS EN 16 SM 0104, Statistics Sweden, Stockholm.

Silveira, S. Ed. (2001) *Building Sustainable Energy Systems – Swedish Experiences*, Swedish Energy Agency.

Swedish Energy Agency (2001) Utveckling av nätavgifter (Development of Net Fees), 1 Jan 1997–2001, Eskilstuna.

Swedish Energy Agency (2002a) Energy in Sweden 2002, Eskilstuna.

Swedish Energy Agency (2002b) Energy in Sweden 2002: Facts and figures 2002, Eskilstuna.

Chapter 9

Cofiring Biomass and Natural Gas – Boosting Power Production from Sugarcane Residues

Arnaldo Walter, Mônica R. Souza[1] *and André Faaij*

9.1. WHY COFIRING?

The term cofiring has been often applied to designate the combined use of fuels in power plants as well as in industrial steam boilers. A special case is the combined use of biomass and fossil fuels, the most acknowledged idea being the mix of biomass and coal in power plants. In some countries, such as the United States, The Netherlands, Austria and Finland, cofiring biomass and fossil fuel has been commercially practiced for power production since the mid-1990s.

Environmental issues, mainly those concerned with mitigation of airborne emissions (carbon dioxide and other gaseous pollutants, especially sulfur oxides), are the main reasons for pursuing efforts on cofiring. Owing to substantial reduction of technical and economic risks, cofiring has also been considered as the first step in enhancing biomass utilization for power generation in some countries. With cofiring, for instance, it is possible to take advantage of the relatively high efficiency of large coal boilers without incurring a large investment (Sondreal et al., 2001).

Cofiring biomass and natural gas has been considered to a less extent so far, and no significant commercial experience has been identified. Recently, a report on cofiring biomass-derived fuels and natural gas in gas turbines has been released in The Netherlands (De Kant and Bodegom, 2000). The research has focused on the technical feasibility and the potential of cofiring low-heating-content fuels and natural gas over different power configurations. Gas turbine constraints and required adaptations have been inventoried with the gas turbine suppliers. Likewise, a similar study was developed some years ago at the National Renewable Energy Laboratory – NREL in the United States, but only a preliminary analysis of technical options was conducted at that time (Spath, 1995).

[1] Monica Rodrigues de Souza is grateful to CNPq and CAPES for the financial support received during her work at University of Campinas – Brazil and at STS, Universiteit Utrecht, The Netherlands.

Considering the biomass use, the term cofiring has been applied in a widespread sense. Strictly, cofiring corresponds to burning a mix of fuels in the same thermal device. However, cofiring has also been understood as: (i) biomass or fossil fuel use to complement the main fuel supply; (ii) biomass use to increase plant capacity, burning fuels without mixing; and even (iii) when biomass is used to full substitution of a fossil fuel in an existing power plant.

This chapter describes a research focused on opportunities for developing power production from sugarcane residues (sugarcane bagasse and sugarcane trash, i.e. leaves and tops of the plant) based on cofiring with natural gas. The three technical alternatives presented are based on the wide definition of cofiring mentioned earlier.

9.2. THE RATIONALE

Brazil is the largest producer of sugarcane in the world. The production in the harvest season of 2000–2001 reached 252 million tons, but it was as large as 315 million tons in the harvest season of 1999–2000. In that period, the amount of bagasse available at the sugar mills reached 780 PJ. Bagasse is inefficiently consumed in the cogeneration systems of sugarcane mills, generating steam that is first used to produce power and, subsequently, to fulfil process thermal demand. In addition, a small amount of the bagasse is traded and used as fuel by other industrial branches, but these market opportunities are constrained by transport costs and the low prices of fuel oil. In addition, this market tends to be further reduced, as natural gas is made available.

Tops and leaves of the sugarcane plant – the so-called sugarcane trash – are currently burned in the field before manual harvesting. Since this practice will be gradually reduced in the next 10–12 years for environmental reasons, it is predicted that the availability of sugarcane residues in Brazil will steadily increase. Sugarcane trash shall be recovered from the fields through mechanic harvesting, a technology that has been introduced in Brazil in the last few years. Potentially the availability of sugarcane trash is as large as bagasse, but topographic constraints will determine how much is economically recoverable (see also Braunbeck et al., Chapter 6).

Based on opportunity costs for sugarcane bagasse and on predicted costs for sugarcane trash recovery, it is foreseen that the cost of this biomass would be lower than 2 US$/GJ and, in some cases, even lower than 1 US$/GJ. Despite the focus given to sugarcane residues in this study, a wide range of biomass could obviously be used for the purpose of cofiring, such as wood chips, bark, thinnings, sawdust, various agriculture residues, etc.

In Brazil, the installed electricity generation capacity is largely based on hydropower. It is estimated that almost 80 per cent of the current capacity (slightly larger than 85 GW) corresponds to hydropower plants. Clearly this enormous dependency on just one energy source is risky and a diversification of power sources is advisable. In fact, in 2001, due to lack of investments in power generation and to a drier summer than usual, power shortages have occurred. Power production from biomass provides both an opportunity for diversification and for expanding the use of renewables in the Brazilian matrix.

Natural gas power plants shall be built in Brazil in the next 5 to 10 years, fostered by governmental policies. Brazilian natural gas reserves are small but the supply is enlarged through imports from Bolivia and, possibly, from Argentina in the near future. As the natural gas market is not yet well developed, thermal power plants have been considered necessary to assure the feasibility of pipeline projects. This would allow the consumption of a large amount of natural gas during the early years of a "take-or-pay" contract.

Additionally, it is important to bear the perspective of private developers in mind, the main investors after privatization and deregulation of the electricity sector. Natural gas thermal power plants appear to be the main option of investment due to the short construction time of the plants, relatively low capital costs ($/kW installed), high efficiency, and large availability. On the other hand, investors identify a risky picture due to the necessity of bulk imports of natural gas and the instability of the Brazilian economy. Medium- to long-term fluctuations of natural gas prices are obviously a matter of concern for investors. The combined use of natural gas with biomass can reduce these risks and increase the fuel flexibility of new power generation capacity. This point is especially relevant in a natural gas market that is still under development.

Two additional points should be observed concerning the natural gas supply. First, the brand new Bolivia–Brazil pipeline crosses – or is relatively close to – the region where approximately 60 to 65 per cent of the Brazilian sugarcane production takes place. Second, as the natural gas market is further developed, and better opportunities for natural gas consumption arise – for instance, on premium markets such as the residential and industrial sectors – biomass could replace natural gas on thermoelectric power plants that shall be built during the early years of the pipeline operation. The period required for the development of a new natural gas market is around ten years.

Competitiveness of electricity production from biomass will strongly depend on the development of new conversion technologies and on the scale of power plants. Future power production from biomass could be based on gasification, for example. Gas turbines are power devices that have some important attributes: reasonable thermal efficiency and initial capital costs that are not as affected by scale effects. It is expected that, with the not-yet-commercially-available BIG-CC cycles (Biomass Gasification Integrated to Combined Cycles), the efficiency of electricity production could reach about 35 to 45 per cent (Walter et al., 2000).

Performance penalties associated with gas turbine adaptation to gasified biomass are meaningful. Biomass-derived gas from air-blown gasifiers has only about 8 to 10 per cent of the energy content of natural gas, resulting in larger mass flow through gas turbines. As a consequence, technical problems can occur, such as compressor surge, increased thermal and mechanical loads on compressor airfoils, the need of an adapted combustion/injection system and problems with flame stability (Rodrigues et al., 2003a). Regarding the BIG-CC technology, cofiring with natural gas is here mainly proposed as a short-term approach to cope with penalties on both efficiency and power resulting from gas turbine derating. The expansion of power plant capacities due to cofiring is also an important contribution for the competitiveness of electricity production from biomass (Rodrigues et al., 2003b).

However, the BIG-CC is not the only technical option for cofired biomass and natural gas plants. Commercial and proven technologies could be used as well. For instance, biomass could be fired independently from natural gas, producing steam in conventional boilers. In addition, steam production would be complemented from HRSGs – heat recovery steam generators – and both streams could be mixed to feed steam turbines of combined cycles. Furthermore, from strategic and economic points of view, electricity production from cofiring natural gas and biomass could be effectively developed as an alternative for the reduction of GHG emissions.

9.3. CASES AND HYPOTHESES FOR SIMULATION

Three cofiring alternatives are analyzed in this study. Case A corresponds to the full displacement of natural gas for biomass in combined cycle power plants. Case B concerns a partial use of biomass, complementing natural gas in combined cycle power plants. Finally, Case C regards the use of biomass only to increase steam production, which is then used in the bottoming cycle (Rankine cycle) of a combined cycle.

All results presented here are based on computational simulations. For the purpose of simulation, some characteristics of the GE PG6101(FA) (a machine developed to operate continuously with low-heating-value fuel) were considered. Only one complementary result regarding Case B is based on GE Frame 7 class gas turbines, for which some characteristics of the GE PG9171(E) were taken.

The BIG-CC system considered in Cases A and B is based on an atmospheric air-blown gasifier. The gasification technology taken into account here – as well as the subsequent low-pressure gas cleaning system – is similar to that proposed by the Swedish company TPS (Termiska Processer AB). Some of the biomass projects that are aimed at the provision of electricity and are under development at present are based on atmospheric gasification technology. BIG-CC systems based on pressurized gasification are also an alternative and possibly would be more feasible for the range of capacities considered here. Atmospheric systems will probably present a lower biomass to electricity conversion efficiency but, on the other hand, in the short term, fewer problems can be expected with syngas production and its cleanup.

BIG-GT technology (generally speaking, gas turbine cycles integrated to biomass gasification) is still under development. The main technological issues in the demonstration of BIG-GT are concerned with (i) scaling-up the gasifier and gas-cleaning technologies and (ii) gas turbine adaptation to low-heating-content fuel. However, the main problems with the current demonstration projects seem to be the initial costs of first generation plants and the difficulties of arranging a reliable fuel supply for the lifetime of the project at a reasonable cost (Walter et al., 2000).

The simulation of BIG-CC systems is based on the schemes and hypothesis presented by Consonni and Larson (1996). Design conditions considered for the bagasse drying result in a moisture content of 15 per cent by weight at the gasifier entrance, the gas being used for drying is the HRSG flue gas. Gasification takes place with air injection, and the syngas leaves the gasifier at about 870°C.

Raw syngas composition was evaluated based on previous results of gasifier simulation performed by ASPEN® – Advanced System for Process Engineering (Walter et al., 1998b). As proxy, it was considered that sugarcane trash has the same ultimate analysis as bagasse. Bagasse and syngas compositions (raw and clean) considered in this study are presented in Table 9.1.

A code was used to evaluate gas turbine off-design performance, i.e. the gas turbine operation with syngas. Details of the procedure can be seen in Walter et al. (1998a). To avoid very high compressor pressure ratio that could dangerously reduce compressor surge margin, gas turbine derating was considered. The rise of pressure ratio under certain limits can be eventually accepted for the compressor of some industrial gas turbines (Corman and Todd, 1993) but, in general, high-pressure ratio imposes unacceptable problems concerned with the increase of shaft torque and thermal loads on airfoils, making this option very aggressive to the equipment (Johnson, 1990). For syngas burning, it was assumed that the maximum GT compressor pressure ratio is 16.4, while its nominal pressure ratio at ISO basis is 14.9.

A common way to derate gas turbines, i.e. to reduce their output power, is through the reduction of the maximum temperature, which is accomplished with the reduction of fuel flow. The term derating is used here to indicate a strategy for gas turbine control to avoid machine operation with a high pressure ratio. In fact,

Table 9.1. Biomass and syngas composition

Biomass composition		Syngas composition	
Bagasse (and trash) ultimate analysis	% Weight – dry basis	Component	Clean % Molecular weight
Carbon	46.3	H_2	16.69
Oxygen	43.3	CO	19.98
Hydrogen	6.4	CO_2	10.49
Nitrogen		CH_4	2.63
Sulfur	<0.1	C_6H_6	0.33
Ash	4.0	H_2O	3.24
LHV[1] [MJ/kg] (Dry basis)	17.5	N_2	46.64
		NH_3	Negligible[2]
LHV [MJ/kg] (50% moisture)	7.84	Tar	Negligible
		LHV [MJ/kg]	5.16

[1] LHV = Lower heating value.
[2] Ammonia and tar should be completely eliminate on the clean-up process.

gas turbine derating is one of the possible strategies that would allow machine conversion from natural gas to syngas. Other strategies are (i) reducing compressor air flow through control of inlet guide vanes – IGVs, (ii) enlarging the expander cross-sectional area in a permanent change, and (iii) promoting blast-air extraction after the compressor. From the viewpoint of the whole system performance, and consequently from the viewpoint of electricity generating costs, derating is the worst solution (Rodrigues et al., 2003a). However, owing to its simplicity, for the first generation of BIG-CC systems, this is most probably the way gas turbines would be converted to syngas firing.

To simplify the modeling procedure, it was considered that steam is produced in an unfired HRSG at just one pressure level, without reheating. This is a reasonable hypothesis for stand-alone BIG-CC cycles due to the requirement of a minimum temperature for HRSG exhaust gases (Consoni and Larson, 1996), but it is not the case for combined cycles burning natural gas. Indeed, it is well known that high efficiency combined cycles require two or three steam pressure levels besides reheating (Bathie, 1996). For instance, for the combined cycle based on PG6101(FA), when natural gas is burned, three steam pressure levels and reheating at the intermediate pressure level is recommended. It is considered that the steam pressure at the turbine entrance is 100 bar. Steam temperature is a function of the GT exhaust gases temperature, while the maximum steam temperature was assumed to be 538°C. Steam is extracted from the steam turbine at 0.48 MPa to feed the deaerator, while the remaining flow is condensed at 9.6 kPa. The temperature of the HRSG feed water is assumed to be constant in all the simulated cases (120°C).

Table 9.2. Case A – Simulation results for natural gas and syngas

Parameter	NG	Syngas
Gas turbine		
GT power [MW]	67	77.5
Pressure ratio	14.9	16.4
Firing temperature [°C]	1288	1178.8
Fuel consumption [kg/s]	4.3	38
Thermal efficiency [%]	32.8	34.7
Steam cycle		
Steam cycle power [MW]	28	31.7
Steam pressure [bar]	100	100
Steam temperature [°C]	538	510.5
CC net power [MW]	94	87
CC net efficiency [%]	46	39[1]

[1] Value corresponds to the overall biomass-to-electricity efficiency.

9.4. SIMULATION AND FEASIBILITY RESULTS

Case A

Simulation results considering combined cycle operation using only natural gas or biomass-derived gas are presented in Table 9.2. A detailed information about this alternative can be found in Walter et al. (1998b). These simulation results correspond to combined cycle operations under the ISO basis. As can be seen, despite derating, the thermal efficiency of a gas turbine operating with syngas is higher than with natural gas, owing to the increase on gas mass flow and to the higher GT compressor pressure ratio. With syngas firing, more power is also produced by the steam bottoming cycle due to the increase of GT exhaust gas flow, but steam is produced at a lower temperature (lower exhaust gas temperature). Albeit more power production both at the gas turbine and at the steam cycle, the system net power is lower when syngas is burned due to high power consumption of plant auxiliaries (mainly the syngas compressor). As mentioned before, the simulation results for natural gas combined cycle are not optimized as it is considered that steam is generated at just one pressure level at the HRSG. A single combined cycle unit (one gas turbine, one HRSG and one steam turbine) of the same capacity can operate with efficiency as high as 53 per cent (Gas Turbine World, 2000).

For current commercial natural gas combined cycles (NG-CC), data of unit capital costs were taken from the literature (Gas Turbine World, 2000). The turnkey unit capital cost of a combined cycle based on PG6101(FA) is estimated at 680 US$/kW. The levelized electricity generating cost was calculated considering the following assumptions: 30-year life, 12 per cent real pretax discount rate, capacity factor 0.85,

O&M (operation and maintenance) costs at 0.4 cents/kWh, and natural gas cost equivalent to 2.5 $/MMBTU. All costs in this study are presented in 1999 US$.

For BIG-CC systems based on atmospheric air-blown gasifiers, the installed investment cost was estimated using Eq. (1). This equation is primarily based on estimates available in the literature for the first commercial plant and incorporates "learning effects" (progress ratio 0.80 and the 5th similar commercial unit). Estimates given by this equation are very close to the estimates presented by other authors for BIG-CC systems of the same net capacity.

$$k_{A\ IGCC} = 5612\ (MW)^{-0.2953}\ [\$/kW] \tag{1}$$

For BIG-CC systems, the nonfuel operation and maintenance cost was estimated at 8.2 US$/MWh. The average cost of biomass (bagasse and trash) was estimated at 8 US$/t. To assure a proper comparison with NG-CC results, some assumptions are common in both cases as, for instance, plant-year life, real discount rate and plant capacity factor.

Estimated levelized electricity generating costs are presented in Table 9.3. The feasibility analysis is based on the evaluation of the internal discount rate (IDR) of the investment. Besides the aforementioned assumptions, it was also considered that (i) the plant construction time is two years in all cases, and 80 per cent of the investment is made during the first year, (ii) taxes are evaluated over the net revenue (as a simplification, a 15 per cent duty was considered), and (iii) the depreciation was calculated along 10 years, using a linear model. Revenues were calculated considering that all electricity can be sold at 45 US$/MWh. Actually, this assumption does not correspond to the reality of the electricity market post deregulation since, in a competitive environment, investors need to define their prices either based on their actual generating costs or on their own profit expectation.

The cofiring case presented in Table 9.3 corresponds to the substitution of fuel, from natural gas to biomass. It was considered that the investment leading to this

Table 9.3. Case A – main results of feasibility analysis

Combined cycles	NG-CC		BIG-CC
Thermal efficiency [%]	46	53	39
Cost of electricity – COE [$/MWh]	36.8	34.1	45.0
IDR (%)	21.4	24.2	12.8
Cofiring	NG to syngas		
Thermal efficiency with NG-CC [%]	46		53
IDR (%)	15.4		16.6
IDR with carbon credit (10 $/t CO_2)	17.3		18.5
IDR with carbon credit (20 $/t CO_2)	19.0		20.2

substitution starts after four years of operation with natural gas. After the sixth year of operation, just syngas is burned at the gas turbine. The investment required for the substitution was estimated as the difference of the installed unit capital cost for both the options.

The feasibility analysis also includes the evaluation of the impact of credits based on avoided emissions of carbon dioxide. These credits could be paid, for instance, by international funds on the context of the Clean Development Mechanism to boost projects aimed at reducing carbon dioxide emissions. It was considered that these credits are free of tax duties. Simplifying the analysis, it was also considered that sugarcane has a nil carbon balance. As can be seen in Table 9.3, each dollar earned per ton of carbon dioxide not released could imply an increment on the IDR of about 0.2 per cent. The results considered two options for the natural gas combined cycle – 46 per cent, that is the simulation result, and 53 per cent, that is the expected thermal efficiency of a single combined cycle unit based on a Frame 6(FA) gas turbine (steam produced at three pressure levels, with reheating).

Case B – preliminary results

The simulation results of a preliminary analysis of Case B are presented in Figure 9.1 for different rates of natural gas replacement. More details of the simulation procedure adopted are presented in Walter et al., 1999. The far left points correspond to gas turbine operation with natural gas only, while the far right points correspond to gas turbine operation just with biomass-derived gas. Gas turbine derating influences combined cycle performance. The gradual switch from natural gas to syngas, as far as GT derating is not necessary, increases power production due to (i) a larger gas

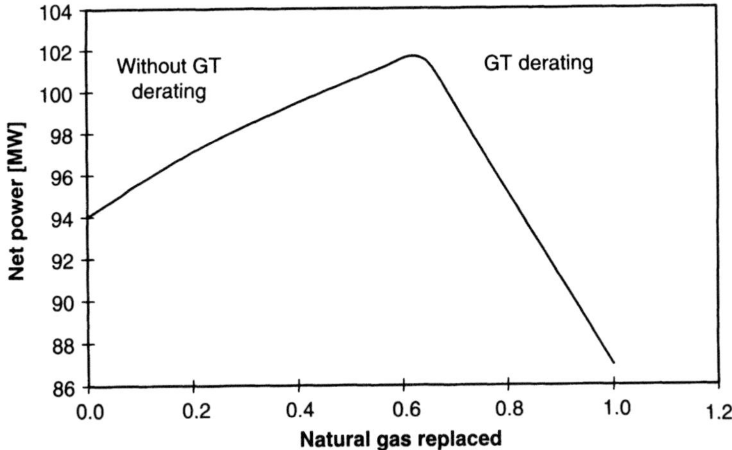

Figure 9.1. Combined cycle power production as a function of natural gas replacement (mass basis).

mass flow and (ii) an increase in compressor pressure ratio. When a larger amount of syngas is used and GT derating is imposed, net power production drops due to derating itself and due to the increasing power consumption of auxiliaries (mainly syngas compressor and air compressor).

Based on the same economic assumptions previously presented for Case A, a feasibility analysis of partial switch from natural gas to syngas was developed. For all intermediary cases between the two limits (GT operation on natural gas and on syngas only), the installed capital cost was estimated as the cost of a conventional natural gas combined cycle plus the capital cost of a biomass gasifier and gas-cleaning unit. The additional investment on syngas production was calculated according to the amount of biomass required (as received, with 50 per cent moisture), assuming a reference value (Faaij et al., 1997) and a scaling factor of 0.70. Finally, a 1.6 factor was applied over the estimated capital cost to take into account the overall set of extra costs regarding GT adaptation and equipment installation.

The cost of electricity as a function of the share of natural gas replaced is presented in Figure 9.2. The same figure includes results that correspond to the impact of credits due to carbon dioxide abatement. As can be seen, carbon credits equivalent to 10 US$/t of CO_2 avoided could make the cofiring option feasible vis-à-vis electricity generation from natural gas in a large range of mixtures between natural gas and biomass-derived gas. At a credit price equivalent to 20 US$/t CO_2 avoided, electricity generation from cofiring natural gas and biomass is even cheaper than some conventional alternatives.

Finally, Figure 9.3 shows the evolution of investment IDR as a function of the share of natural gas replaced and carbon credits as well. The decline of IDR as far as gas turbine operation is switched from natural gas to syngas is due to the higher

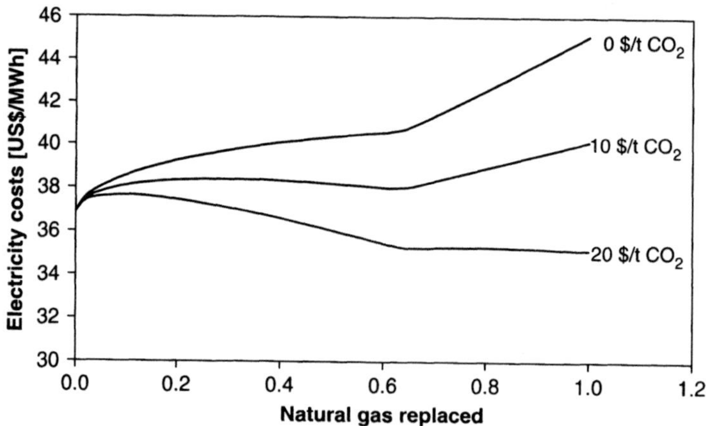

Figure 9.2. Estimated electricity costs as a function of natural gas replacement and carbon credits derived from CO_2 abatement.

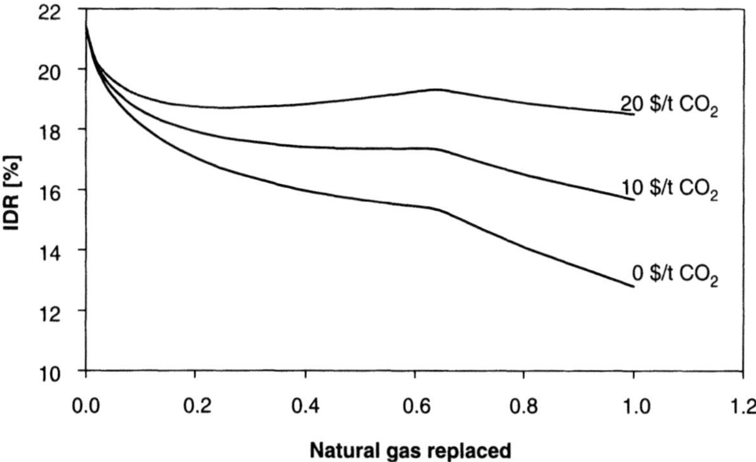

Figure 9.3. IDR as a function of the share of natural gas replaced and use of carbon credits.

cost of electricity (0 \$/t CO_2). Carbons credits make IDR reduction smoother or can counterbalance this tendency at higher carbon credit prices.

Case C

Case C was defined considering a natural gas combined cycle plant based on two PG6101(FA) units, one unfired HRSG for each gas turbine and just one steam turbine. The total capacity of the NG-CC is 187.2 MW, with an additional contribution of 53.2 MW from biomass. Thus, the total plant capacity is 240.4 MW. The steam cycle parameters and the overall cycle performance are the same as presented in Table 9.2. The biomass contribution to the cofiring plant occurs in a conventional steam power cycle in which steam is generated at the same live-steam parameters as in the HRSGs, producing 53.2 MW as well. Main simulation results are presented in Table 9.4.

The feasibility assessment was accomplished based on the assumptions described earlier for a NG-CC power plant. The initial capital cost for a 185 MW class was estimated as 480 US\$/kW. For the steam cycle plant based on biomass, the unit capital cost is estimated at 1110 US\$/kW. This evaluation is based on cost functions for the main plant components and results were checked with published data. Annual operation and maintenance costs (excluding fuel) were considered as equivalent to 5 per cent of the initial capital cost.

The main results of the feasibility analysis are presented in Table 9.5. The addition of a biomass module represents an increase in the plant capacity of about 28 per cent vis-à-vis the natural gas combined cycle plant. As the cost of producing electricity

Table 9.4. Case C – Main simulation results

Gas turbine power [MW]	2 × 67
Conventional steam CC bottoming cycle [MW]	53.2
Steam cycle based on biomass [MW]	53.2
Total capacity [MW]	240.4
Natural gas consumption [kg/s]	8.6
Biomass consumption [kg/s]	25.93
NG-CC cycle efficiency [%]	46.0
Biomass plant efficiency [%]	26.2
Overall plant efficiency [%]	41.6

Table 9.5. Case C – Results of the feasibility analysis

Investment on biomass part [\$/kW]	NG-CC	Cofiring 1100	Cofiring 500
Cost of electricity [\$/MWh]	33.5	35.3	31.9
IDR (%)	28.7	23.3	30.9
IDR with carbon credit (10 \$/t CO_2)	–	24.6	32.6
IDR with carbon credit (20 \$/t CO_2)	–	25.9	34.3

through the biomass plant running all over the year (capacity factor 0.85, as well as for the NG-CC) is higher than the cost estimated for electricity produced in the NG-CC plant (40.0 × 33.5 US\$/MWh), there is a rise in the cost of electricity produced by the cofired plant.

For full-year operation of the biomass-based power plant, the IDR is significantly reduced vis-à-vis the NG-CC case. The investment would be more feasible if carbon credits were available but these should be as high as 40 \$/t CO_2 to make the cost of electricity of the cofired plant equivalent to the NG-CC.

However, if the biomass-based power plant operates for a shorter period during the year (e.g. just during the 6–7 months of the sugarcane harvest season), the final cost of electricity rises. For instance, for 4080 h of operation per year (200 days of harvesting and a capacity factor of 0.85 during this period), the cost of electricity produced by the biomass power plant rises to 61.4 US\$/MWh and the cost of electricity of the cofired unit increases to 37.7 US\$/MWh. Simplifying the analysis, it was considered that the average biomass cost is 8 US\$/t in both cases. The IDR is estimated as 19.9, 20.6 and 21.4 per cent for the three cases considered (depending on the value of the carbon credits).

The Case C was also evaluated considering a capital cost for the steam cycle plant based on biomass at 500 US\$/kW. This is a reasonable estimate for equipment built in Brazil as the cost in US dollars has dropped significantly due to the devaluation of Brazilian currency since early 1999. As can be seen in Table 9.5, in this case, the cost of electricity generated in the cofired unit could be lower than in the natural gas combined cycle unit, even without carbon credits.

9.5. COMPARISON OF ALTERNATIVES

A combined cycle plant based on a PG6101(FA) gas turbine, such as the one considered in this study, and following the hypothesis assumed, consumes about 500 thousand Nm^3/day of natural gas operating at full load basis (about 430 Nm^3/day for an average capacity factor of 0.85). Thus to guarantee full operation of a 100 MW plant, some 540 thousand Nm^3/day would be needed. Table 9.6 summarizes the amount of natural gas that can be displaced per set of 100 MW power capacity for each cofiring alternative analyzed in this study. Results for Case B correspond to the best case from the point of view of capacity, i.e. 101.7 MW of net capacity, for 61.9 per cent substitution of natural gas on mass basis.

To give a perspective of this contribution, we look into the present situation in Brazil. The Brazilian government has recently decided to support the installation of almost 17 GW of thermal power capacity, totalling 49 plants, most of them designed to fire natural gas. Of this total capacity, 6.5 GW are planned in industrialized sites of the State of São Paulo, not far from the regions where there is a large concentration of sugarcane mills. The natural gas demand to allow the operation of 6500 MW new capacity in the State of São Paulo is estimated at 25 to 26 million Nm^3/day, taking into account the predicted average annual capacity factor and actual natural gas to electricity efficiencies (in round numbers, larger than the performance figures used in this article). This natural gas volume is quite substantial as a share of the capacity of the brand new Brazil–Bolivia natural gas pipeline, the main source of natural gas supply for the years to come.

The adoption of a Case C strategy would allow a maximum displacement of natural gas of about (124.6/577.0) × 25–26 million Nm^3/day = 5.4–5.6 million Nm^3/day. With the strategy that corresponds to Case B, the maximum volume of natural gas that could be displaced would be (345.6/577.0) × 25–26 million Nm^3/day = 15.0–15.6 million Nm^3/day. On the other hand, theoretically, a Case A

Table 9.6. Comparison of Case A, Case B and Case C

Parameter	NG-CC	Case A	Case B	Case C
NG demand [Nm^3/MWh]	222.4	–	78.4	170.5
NG displaced [1000 Nm^3/day/ 100 MW]	–	577.0	345.6	124.6
Biomass demand [t/MWh]	–	1.16	0.7	0.39
[t/day/100 MW]	–	2788.8	1682.4	926.6

strategy would allow 100 per cent natural gas displacement. It is clear from the numbers presented that cofiring biomass and natural gas could make the natural gas market more flexible, without drawbacks to electricity generation.

Sugarcane production in the State of São Paulo is estimated at 180 million tons per year. With this level of production, some 23.2 million tons of sugarcane trash could be recovered (50 per cent of the amount available, owing to topographic constraints; sugarcane trash availability is equivalent to approximately 25 per cent of sugarcane mass). The full availability of bagasse is estimated at 46.8 million tons/year (13 per cent fiber content and 50 per cent moisture), the ordinary surplus being about 10–15 per cent of the total availability. Achieving 6500 MW capacity, full implementation of the Case C strategy would demand 18.7 million tons of biomass per year, while the Case B strategy would require 33.9 million tons/year. Nowadays, trash is essentially burned at the field before sugarcane harvesting. For bagasse, there is not a market able to consume all the existing surplus. Hence, from the point of view of biomass availability, there is no particular constraint regarding the implementation of the cofiring alternatives analyzed here.

From the economic point of view, the preliminary results indicate that Case B can be considered as a reasonable alternative as the cost of electricity produced is kept at an acceptable level, and the investment IDR is not reduced to any large extent when biomass contribution is considered. If the biomass plant can be built mainly with equipment manufactured in Brazil (thus at lower capital cost), Case C is the best option, as it allows a reduction on the cost of electricity produced while enhancing the investment IDR. According to the results, to substitute biomass for natural gas in BIG-CC power plants (Case A) is not a good alternative as some amount of natural gas allows improvements in the system economics. Obviously, the opportunity to generate carbon credits from cofired plants would make a substantial difference on the economics of such alternatives.

The feasibility of Case B and Case C could be further improved with larger power units, taking advantage of economies of scale and, consequently, reducing the capital costs per unit energy generated. Strictly speaking, the location of the power plant would be a matter of concern regarding biomass transportation costs. The same is true regarding plant size – as the scale of the plant increases, costs related to logistics also increase.

9.6. FINAL REMARKS

Electricity production with the use of a renewable energy source on a sustainable basis is clearly the most important contribution of the cofiring alternatives discussed here. Additionally, natural gas could be used in a more rational manner, allowing gradual changes within the energy matrix and reduction of environmental impacts. In fact, a transition into biomass for power generation can be interesting for natural gas distribution companies. These companies could maximize their take-off in the beginning of the cash flow project and further extend their supply of natural gas at a very low cost (or even no cost) as the consumer market is further developed. This transition into biomass could also be considered for countries where the market of natural gas is already well developed. Another interesting point to be considered concerns the required investment to find new reserves at high marginal costs.

One of the main conclusions of a detailed study of the cofired option presented in Case B was that cofiring could be instrumental for the market development of the BIG-CC technology in its early stages. In fact, besides the possibility of overcoming technical constraints for the very first BIG-CC units, the enlargement of the fuel heating value due to the natural gas contribution boosts the plant efficiency and, consequently, contributes to the reduction of the cost of electricity generated from biomass. The cost reduction is, to a large extent, due to economies of scale. Small biomass-fueled gasification plants can benefit from the high efficiency and low capital costs of large combined cycles without scaling up the biomass parts. However, in the medium- to long-term, as the BIG-CC is further developed and gas turbines for biomass-derived gas are redesigned, cofired combined cycles would probably be less justifiable (Souza, 2001). In the case of Brazil, successful experiences with cofiring natural gas and biomass could not only help foster BIG-CC market expansion, but also could allow a more efficient use of sugarcane residues for electricity generation.

Measures

kPa = 10^3 N/m^2

LHV = Lower heating value

MMBTU = Million of British Thermal Units

Mol = Molecular weight

MPa = 10^6 N/m^2

REFERENCES

Bathie, W. (1996) *Fundamentals of Gas Turbines*, 2nd Edition, John Wiley & Sons, New York.

Consonni, S. & Larson, E.D. (1996) Biomass-gasifier/aeroderivative gas turbine combined cycles: part B - performance calculations and economic assessment in *Journal of Engineering for Gas Turbines and Power*, **Vol. 118**(3), pp 516–525.

Corman J.C. & Todd, D.M. (1993) Technology considerations for optimising IGCC plant performance, *Proceedings of International Gas Turbine and Aero Engine Congress and Exposition*, ASME, paper 93-GT-358, Cincinnati, Ohio, USA.

De Kant, H.F. & Bodegom, M. (2000) Study on Applying Gasifiers for Co-firing Natural Gas Fired Energy Conversion Facilities, NOVEM, The Netherlands (In Dutch).

Faaij, A., van Ree, R., Waldheim, L., Olsson, E., Oudhuis, A., van Wijk, A., Daey-Ouwens, C. & Turkenburg, W. (1997) Gasification of biomass wastes and residues for electricity production in *Biomass & Bioenergy*, **Vol. 12**(6), Elsevier Science, pp 387–407.

Gas Turbine World (2000) *Gas Turbine World 1999–2000 Handbook*, Pequot Publishing Inc., V. 20, Fairfield, CT, USA.

Johnson, M.S. (1990) The Effects of Gas Turbine Characteristics on Integrated Gasification Combined – Cycle Power Plant Performance, PhD dissertation, Stanford University.

Rodrigues, M., Walter, A. & Faaij, A. (2003a) Performance Evaluation of Atmospheric BIG-CC Systems Under Different Gas Turbine Control Strategies, Paper submitted to *Applied Energy*, Elsevier Science.

Rodrigues, M., Faaij, A. & Walter, A. (2003b) Techno-Economic Analysis of Co-fired Biomass Integrated Gasification/Combined Cycle Systems with Inclusion of Economies of Scale in *Energy* **Vol. 28**(12), Elsevier, pp 1229–1258.

Sondreal, E.A., Benson, S.A., Hurley, J.P., Mann, M.D., Pavlish, J.H., Swanson, M.L., Weber, G.F. & Zygarlicke, C.J. (2001) Review of advances in combustion technology and biomass cofiring in *Fuel Processing Technology* **Vol. 71**(1–3), Elsevier, pp 7–38.

Souza, M.R. (2001) Cofiring as a Tool to Booster Biomass Integrated Gasification/Combined Cycle Technology, PhD dissertation, State University of Campinas, Campinas, Brazil (in Portuguese).

Spath, P. (1995) Innovative Ways of Utilizing Biomass in a Cofiring Scenario with a Gas Turbine Integrated Combined Cycle System, Biomass Power Milestone Completion Report, National Renewable Energy Laboratory, Golden, CO, USA.

Walter, A., Faaij, A. & Bauen, A. (2000) New technologies for modern biomass energy carriers, in *Industrial Uses of Biomass Energy – the Example of Brazil*, Eds. Rosillo-Calle, F., Bajay, S.V. & Rothman, H., Taylor & Francis, London, pp 217–228.

Walter, A., Souza, M.R. & Overend, R.P. (1999) Feasibility of cofiring (biomass + natural gas) power systems, *Proceedings of the Fourth Biomass Conference of the Americas*, August, Oakland, CA, USA, pp 1321–1327.

Walter, A., Llagostera, J. & Gallo, W. (1998a) Impact of gas turbine derating on the performance and on the economics of BIG-GT cycles, *Proceedings of ASME Advanced Energy Systems Division – 1998*, ASME International Mechanical Engineering Congress and Exposition, November, Anaheim, CA, USA, pp 67–72.

Walter, A., Souza, M.R. & Overend, R.P. (1998b) Preliminary evaluation of cofiring natural gas + biomass in Brazil, *Proceedings of the Brazilian Meeting on Thermal Sciences – ENCIT 98*, November, Rio de Janeiro, Brazil, pp 1220–1226.

Chapter 10

Techno-Economic Feasibility of Biomass-based Electricity Generation in Sri Lanka

Priyantha Wijayatunga, Upali Daranagama and K.P. Ariyadasa[1]

10.1. INTRODUCTION

Biomass accounts for 51 per cent of energy supply in Sri Lanka (see Figure 10.1). Most of this biomass-based energy use is traditionally confined to the domestic sector, where most of the rural and suburban households rely on fuelwood for cooking. Industrial and agricultural sectors also use wood fuel as well as other biomass-based material such as bagasse and rice husk, to generate heat or steam for agricultural processes and to drive small-scale industrial processes (Ceylon Electricity Board, 1998).

In recent times, biomass has attracted widespread interest as a primary energy source for electricity generation, due to its potential as a low cost, indigenous supply of energy as well as due to environmental benefits accompanying biomass-based generation technology. Conversion of biomass to electricity is considered as one option available to arrest CO_2 emissions caused by fossil-fuel-based generation. In addition to this global benefit, there are local benefits mainly resulting from energy plantations accompanying biomass-based generation technology, such as reduced soil erosion, restoration of degraded lands, and amelioration of local impacts of fossil-fired power generation (e.g. SO_x and NO_x). Other advantages include social benefits such as creation of local employment and improved availability of fuelwood for household use.

10.2. LAND AVAILABILITY

Potential land area

Biomass-based electricity generation and its environmental appeal largely depend on the availability of land to set up energy plantations to satisfy the fuel requirement

[1] The authors wish to express their gratitude to the Energy Forum, Sri Lanka for its financial assistance to carry out this study. The help extended by the Department of Survey, Department of Forests, Sri Lanka and the Energy Conservation Fund of Sri Lanka in providing necessary data is also gratefully acknowledged.

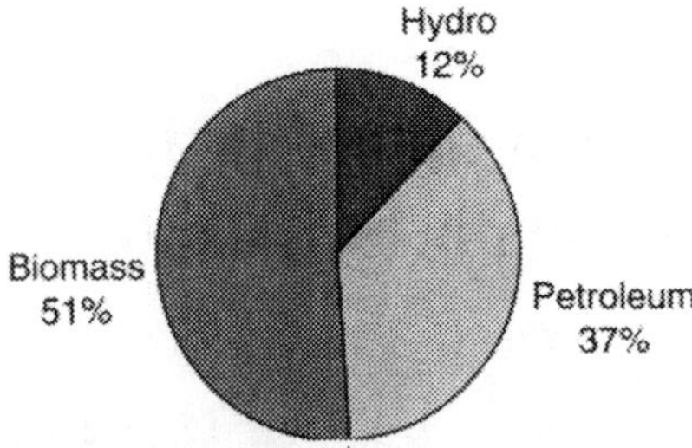

Figure 10.1. Primary energy supply in Sri Lanka (1999).

Table 10.1. Distribution of land by type of landscape 1987

Type	Area (ha)
Urban land	29 190
Agricultural land	3 714 110
Forest land	1 761 360
Range land	593 520
Wet land	60 290
Water	290 520
Barren land	77 480

of the power plants. The total land area of Sri Lanka amounting to over 6.5 million hectares (ha) can be categorized into Urban land, Agricultural land, Forest land, Range land, Wet land, Water bodies and Barren land (see Table 10.1) (Jayasinghe, 1998).

Agricultural land usage includes sparsely used croplands, which accounts for nearly 1.3 million hectares. The agricultural activities in Sri Lanka can be divided into lowland cultivation and upland cultivation. Lowland cultivation mainly consists of paddy cultivation whereas upland cultivation is mainly in the form of dry farming. The chena or shifting cultivation is the main form of dry farming, covering an estimated area of 1 million hectares. The nature of this type of cultivation is such that the utilization of land is cyclic and, therefore, the total area occupied by these activities is underutilized at any given time. Though the extent of land under-utilization has not yet been properly evaluated, several socioeconomic studies have revealed that the livelihood of the rural farming community occupying these lands is mainly dependent on the agricultural activity (Ariyadasa, 1996).

The maximum area available for energy plantations is what has been identified as sparsely used cropland and scrubland, amounting to a total of approximately 1.7 million hectares in all districts. It is important to note that some of the areas under scrubland and sparsely used cropland have very steep terrain for properly

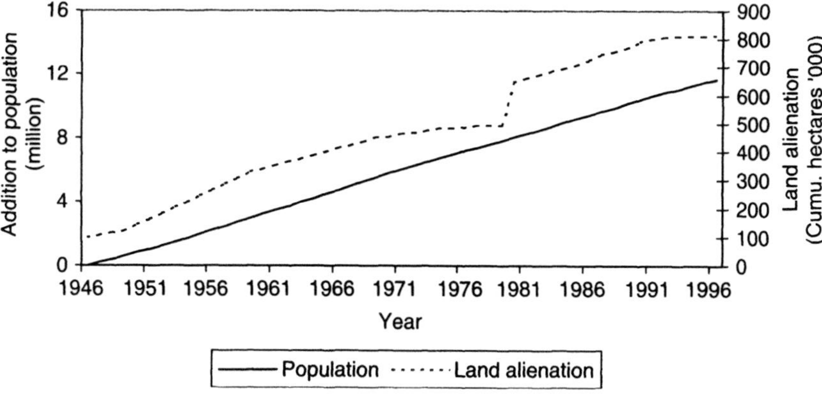

Figure 10.2. Variation of addition to the population and cumulative land alienation by the State, 1946–1996.

managed energy plantations due to many potential adverse environmental impacts such as soil erosion associated with felling and transport of trees in such areas.

Land alienation

With the increase in the country's population and the improvement in the standard of living as a result of economic development, new claims for residential land and land for other economic activities is inevitable. The total area covered by homesteads has increased by 0.8 million ha between 1939 and 1998 mainly as a result of village expansion and resettlement schemes and encroachments into state land (see also Figure 10.2) (Government of Sri Lanka, Land Commissioners Department, 1946–1996). Such increased usage of land for homesteads is expected to proceed.

This shows that there has been a strong correlation between the population growth and land alienation by the state. It can be seen that for each person added to the population, approximately 0.07 ha of land area is to be set aside. The population in the country is expected to stabilize at 25 million by year 2025 with around 6.5 million persons added from 1998. This results in an additional land requirement of around 450 000 ha over the next 25 years.

Maximum usable land

As discussed earlier a minimum of 450 000 ha of land need to be set aside for homesteads during the next 25 years to meet the population growth until it stabilizes. This results in a drop of maximum land availability for energy plantations to 1.2 million hectares. In addition, increasing economic activities including expansion of the agricultural sector and restricted water availability for plantation establishment make it necessary to exclude a considerable extent of land out of what is identified as the maximum available land area for energy plantations. Only then can one estimate the actual availability of fuelwood for possible biomass-based electricity generation in Sri Lanka.

10.3. ENERGY PLANTATIONS IN SRI LANKA

Considering factors such as the local climate and soils, a number of species have been identified as appropriate for available lands in Sri Lanka. These species offer a high degree of certainty to produce firewood for biomass-based thermal power plants (see Table 10.2).

Except the Eucalyptus species mentioned in Table 10.2, all others are legumes that enrich soils by fixing atmospheric nitrogen. This also indicates their ability to grow in poor sites, and lower the requirements for fertilizers. Therefore, these species are very much suited for degraded sites found in the dry and intermediate zones in Sri Lanka (Evans, 1992; Ariyadasa, 1996; FAO, 1993, 1997).

Biomass production from different species are shown in Table 10.3. Although commercial-scale short-rotation energy plantations are rare in Sri Lanka, the growth parameters of the existing fuelwood plantations could be used to determine the potential biomass yields of the recommended species in future short-rotation plantations. The yield-estimates given in Table 10.3 are based on actual field data collected from different plantations situated in different parts of the country.

Table 10.2. Species for energy plantations in Sri Lanka

Wet zone	Intermediate zone	Dry zone
Acacia mangium	*Leucaena leucocephala*	*Leucaena leucocephala*
Leucaena leucocephala (Ipil-Ipil)	*Calliandra calothyrsus*	*Eucalyptus camaldulensis*
Eucalyptus grandis	*Gliricidia sepium*	*Acacia auriculiformis*
Eucalyptus robusta		
Calliandra calothyrsus		
Gliricidia sepium		

Table 10.3. Woody biomass of different species in Sri Lanka

Species	Climatic zone	Biomass production (dry tons/ha/year)
Eucalyptus camaldulensis	Dry zone	10–12
Calliandra calothyrsus	Intermediate/Wet zones	6–8
Gliricidia sepium	Intermediate/Wet zones	12
Acacia auriculiformis	Dry/Intermediate zones	8–10
Eucalyptus grandis	Up country	10

Table 10.4. Average biomass production in each harvesting period

Harvesting period	Yield (tons/ha)
1st after 5 years	50
2nd after 8 years	36
3rd after 11 years	33
4th after 14 years	30

Biomass production of *Eucalyptus robusta* is similar to that of *Eucalyptus grandis* while *Acacia mangium* is similar to *Acacia auruculiformis* (Perlack et al.). *Leucaena leucocephalla* would produce about 8–10 dry tons/ha under Sri Lankan conditions. The average yield during different harvesting periods in a newly established energy plantation in Sri Lanka are given in Table 10.4 (Gunaratne and Heenkenda, 1993).

This gives an annual average yield of 10 dry tons per hectare, with a total of about 7.5 MWh of annual energy per hectare at an overall plant efficiency of 18 per cent, and a 3-year rotation of the fuelwood plantation.

10.4. TECHNOLOGY OPTIONS

The technological options available can be broadly categorized into two main groups, conventional steam cycle based plants and those based on wood gasification technology. The conventional steam cycle power plants are based on proven technology refined over several decades. The overall plant efficiency of the system is in the range of 18–22 per cent. Typical capacities offered with this conventional technology vary between 10 kW and a few hundreds of MW.

There are a number of technological options available to employ air gasification of wood. Down draft fixed bed gasifier, feeding a diesel engine was considered for small-scale power plants ranging from 100 kW. The overall plant efficiency of this system is around 22 per cent. For the range 5–10 MW, two technical options, i.e. pressurized fluidized bed gasification/steam injected gas turbine and pressurized fluidized bed gasification/diesel engine combinations are available. The overall plant efficiencies of these systems are approximately 29 and 34 per cent respectively. For large scale applications it is suitable to use the pressurized fluidized bed gasifier feeding a steam injected gas turbine where the overall plant efficiency can be around 30 per cent (Sipila, 1996; Solantasta, 1995).

Although it offers a high overall efficiency in the range of 30 per cent or above for medium-sized power plants (MW scale), wood gasification-based technology used in biomass-based plants is relatively new and only few plants are in operation on a commercial scale in this region. Therefore, considering the technical feasibility within Sri Lanka, plants operating on conventional steam cycle technology are found to be more appropriate for biomass-based electricity generation systems in the country.

10.5. ECONOMIC ANALYSIS

Plant installation cost

Capital costs of biomass-based power plants vary widely from around 1000 US$/kW to 4000 US$/kW depending on the technology used for electricity generation. The calculation of the installation-related components of the specific energy cost of a typical power plant is given here.

Economic parameters[2]	
Real (Economic) discount rate	10 per cent
Plant economic life	20 years
Escalation	Neglected

Calculation of specific cost	
Investment required	1500 US$/kW
Annual Plant Factor	60 per cent
Energy output (in a 1 kW plant)	$= 8760 \times 0.6 = 5256$ kWh
Present Worth Factor	= 8.51 (10% discount rate for 20 years)
Present Value of energy output	$= 8.51 \times 5256 = 44\,728$ kWh
Specific Cost component associated with power plant installation	= 1500/44 728 = 3.35 US cents/kWh

Fuel and maintenance cost

It is estimated that a land area of 0.5–1 ha is required per kilowatt of power plant capacity, with the power plant operating at an annual plant factor of 60 per cent. Assuming that the power plant operates at an average overall efficiency of 20 per cent, 1.2 kg of fuelwood is required to generate 1 kWh.

Cost of fuelwood	
Cost of fuelwood at plantation	= 113.4 US$/ton
Cost of processing	= 12 US$/ton
Transport	= 4 US$/ton
Total	= 29.4 US$/ton
Rent for land	= 32.3 US$/ha.year
Equivalent rent cost	= 3.0 US$/ton of fuelwood

[2] Inflation is not considered as the analysis is in constant currency. Taxes and duties are neglected in the economic analysis.

Total cost of fuelwood = 32.4 US$/ton
Cost of fuelwood = 3.88 US cents/kWh (at a consumption rate of 1.2 kg/kWh of electricity generated)

Specific cost

The average specific cost of energy from the biomass-based generation plant considered here can be finally determined as:

Specific Cost associated with plant installation = 3.35 US cents/kWh
Maintenance cost of power plant (assumed 10% of above) = 12 US cents/kWh
Cost of fuelwood = 4 US cents/kWh
Total Specific Cost = 29.4 US cents/kWh

The contributions to the final unit cost from different cost components are shown in Table 10.5. It can be seen that the major components of the unit energy cost are those associated with plant installation and felling and processing.

Table 10.5. Cost factors of biomass-based energy cost

Cost component	Contribution to life cycle cost (%)
Plant installation	44.1
Plant maintenance cost	4.5
Rent for land	4.9
Plantation	21.2
Felling and processing	19.0
Transport	6.3

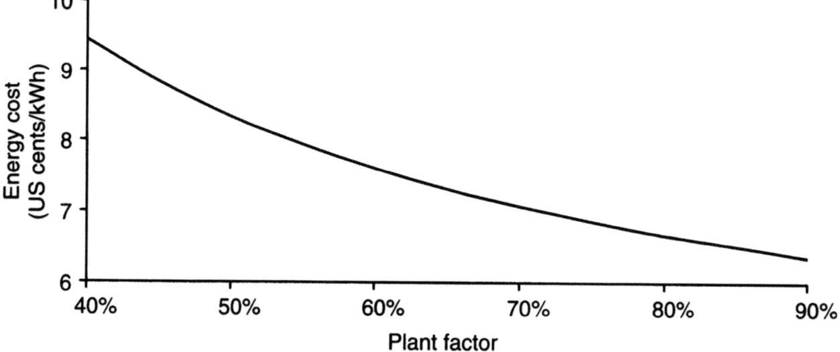

Figure 10.3. Variation of specific energy cost with annual plant factors.

The variation of specific energy cost with plant factors at a typical installation cost of US$ 1500 per kW, is given in Figure 10.3. It can be seen that even at an optimistic annual plant factor of 80 per cent, the specific energy cost is as high as 7 US cents/kWh. To maintain this plant factor, a 100 kW plant needs an energy plantation of 50 ha, continuously being planted, maintained and harvested.

It is generally accepted that renewable energy sources become more attractive in comparison to traditional means of electricity generation, when lower discount rates are used for economic analysis. The variation of specific energy cost with the discount rate is shown in Figure 10.4. Even at a discount rate as low as 2 per cent, the specific energy cost tends to be as high as 5.3 US cents/kWh. This is mainly because of relatively high operational costs associated with biomass-based power plants compared with other forms of renewable energy based electricity generating systems.

When considering possible involvement of the private sector in the development of biomass-based electricity generation, it is important to investigate the effect of the

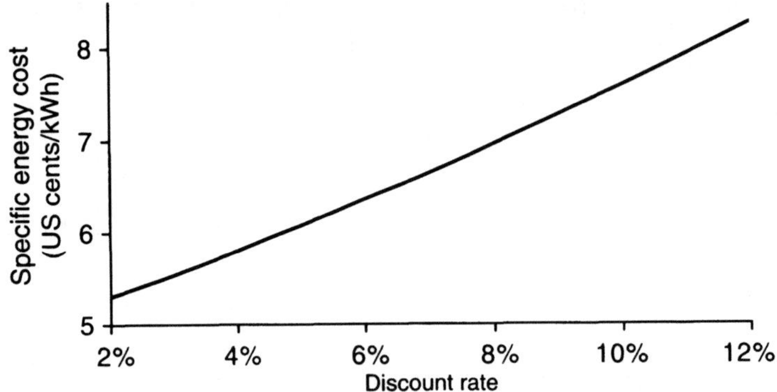

Figure 10.4. Variation of specific cost against discount rate used for economic analysis.

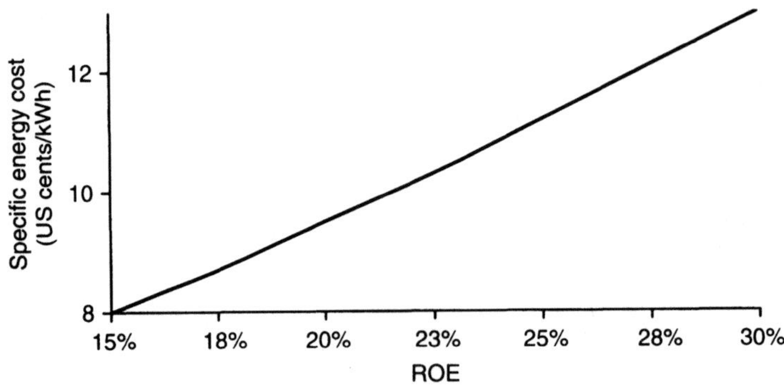

Figure 10.5. Variation of power purchase price with expected IRR on equity.

power purchase tariff of the utility on the Internal Rate of Return on Equity (ROE) of the investor. Figure 10.5 shows the variation of power purchase price to be agreed with the utility for different ROE expectations of the developer of a biomass-based plant.

In Figure 10.5, it is assumed that the plant is offered with a grant component of approximately 30 per cent, while the remaining 70 per cent comes from the developer as equity. The plant is assumed to be operating at a plant factor of 60 per cent. Private sector developers usually expect an ROE of at least 20 per cent and such ROE expectations result in a required minimum power purchase price of approximately 9.5 US cents/kWh.

10.6. CONCLUSIONS

Land availability

Plantation establishment cost in barren land tends to be very high due to bad soil conditions. Therefore, only sparsely used cropland, scrubland and grassland can be used for energy plantations in Sri Lanka. Land at high altitudes with steep terrain needs to be excluded due to complexities in plantation management. This limits the available land potential to the districts of Anuradhapura, Polonnaruwa, Ampara, Badulla and Monaragala. Also, it is recorded that almost all sparsely used cropland are on chena cultivation, providing basic food and other requirements to many rural families. Therfore, in effect, the actual national potential of land available for energy plantations is limited.

Fuelwood production

An array of species, that are well suited to the conditions of available lands, can be successfully used for fuelwood production. Even under low input/low output system of management, these species are capable of producing at least 10 dry tons of fuel wood per hectare per year. With the intensive plantation management, which is often associated with short rotation energy plantations, there is room for considerable improvement in the level of fuelwood production.

Technology

Even if they have relatively low overall efficiencies, generation plants operating on conventional steam cycle are based on proven technology, refined over several decades. It is also important to note that there are many well-established manufacturers throughout the world using this technology including some within the South Asian region. Typical capacities offered with this conventional technology vary between 100 kW and a few hundreds of MW. Reliability of these plants tends to be relatively high and the technical know-how is widespread among many users. Spare parts are widely available as the technology is well established. Considering these factors, plants operating on conventional steam cycle technology seem to be more appropriate for Sri Lanka.

Economics

According to the economic analysis presented in the report, the energy costs associated with biomass-based plants operating on conventional steam cycle technology are comparable with those of large conventional gas turbine plants presently operating on auto diesel. With declining capital costs of biomass-based plants, particularly those from manufacturers in the South Asia region the overall costs are likely to come down in the near future. As a result, biomass-based electricity generation can become competitive in the Sri Lanka electricity generation system where fossil-fuel-based generation has become a significant component.

REFERENCES

Ariyadasa, K.P. (1996) Sri Lanka Profile. In Asia-Pacific Agroforestry Profiles: Second Edition, APAN Field Document No.4 and RAP Publication 1996/20, Asia Pacific Agroforestry Network, Bogor, Indonesia and Food and Agriculture Organization of the United Nations, Bangkok, Thailand.

Ceylon Electricity Board (1998) Long Term Generation Expansion Planning Studies, Generation Planning Branch.

Evans, J. (1992) *Plantation Forestry in the Tropics*, 2nd Edition, Clarendon Press, Oxford.

Food and Agriculture Organization, (1995/96) *Proceedings of the Regional Expert Consultation on Eucalyptus-October 1993*, RAPA Publications.

Food and Agriculture Organization (1997) Development in Agroforestry Research, Los Banos, Laguna, Philippines.

Government of Sri Lanka, Land Commissioners Department Annual Reports (1946–1996).

Gunaratne, W.D.L. & Heenkende, A.P. (1993) Effect of different pruning intervals on biomass production of Gliricidia sepium and yield of Piper nigrum L, *Proceedings of Fourth Regional Workshop on Multipurpose Trees*, Kandy, Sri Lanka, March.

Jayasinghe, J. (1998) Status of Land Resources in Sri Lanka, Personal Communication, October.

Perlack, R.D., Wright, L.L. & Huston, M. (1997) Biomass Fuel from Woody Crops for Electric Power, Bioenergy Systems Report, Winrock International-Biomass Energy Systems and Development Project, USA.

Sipila, K.M. (1996) Power Production from biomass with special emphasis on gasification and pyrolysis, *VTT Symposium 164*, Technical Research Center of Finland.

Solantasta, Y.E. (1995) Feasibility of Electricity Production from Biomass by Gasification Systems, Technical Research Centre of Finland, ESPOO.

Survey Department (1982–1987) CRS Maps on Land use.

Survey Department (1988) The National Atlas of Sri Lanka.

World Bank (1996) Sri Lanka Non Plantation Crop Sector Policy Alternatives, Agriculture and Natural Resources Division.

World Bank (1985) Technical and Cost Characteristics of Dendro Thermal Power System, Energy Department Paper No. 31.

Chapter 11
Classification of Solid Biofuels as a Tool for Market Development

Daniela Thrän, Marlies Härdtlein and Martin Kaltschmitt

11.1. THE NEED FOR A SOLID BIOFUEL STANDARDIZATION

There is a significant potential for an increased use of biomass all over the European Union. Solid biofuels can contribute significantly to reach the political goals of the European Commission and national governments to increase the share of renewable energy and reduce CO_2 emissions from anthropogenic sources. For various reasons, however, this is not happening easily. In many cases, the costs of energy provision are higher for biofuels than for fossil fuels so that additional development programs are urgently needed if this potential is to be rationally explored.

Table 11.1 shows the current use of biomass resources for electricity and heat generation in the EU and the estimated potential from different sources in each country. Large differences can be observed both in the amount and the type of resources being exploited in each country. It can be noted, for instance, that while some countries such as Denmark, the Netherlands and Austria have already exploited 40 to 50 per cent of their potential, Germany is only using 10 per cent of its total potential.

Differences can also be observed in the conversion forms through which biomass resources are being exploited. The use of biomass for district heating has reached quite significant levels in a few countries such as Austria, Finland and Sweden, where mainly fuelwood and wood residues from forestry and wood-processing industries are being utilized. In Germany, woody residues have been used at a more or less steady level in the last ten years for domestic heating purposes, while other biomass resources have not yet been much explored. There is, for example, a significant amount of herbaceous residues, mainly straw, which can be used with technologies that are readily available.

To develop a more widespread use of the solid biofuel resources, the costs of production, provision and use of biomass fuels have to be reduced significantly so that they can compete with fossil fuels economically. With this in mind, it is necessary to consider the possibilities of cost reduction all along the supply chain of

Bioenergy – Realizing the Potential

Table 11.1. Use and potential of biomass in the EU

	Currents use			Potential			
	Electricity	Heat	Total	Woody residues	Herbaceous residues	Energy crops	Total
	(in PJ/yr)			(in PJ/yr)			
Austria	15.6	111.4	127	164.5	22.4	62.7	249.7
Belgium & Lux.	6.7	10.5	17.3	54.4	12.9	37.5	104.8
Denmark	31.1	23.7	54.8	29.2	45.7	60.2	135
Finland	51	154.1	205.1	494	18.5	34.3	546.9
France	38.4	371.2	409.5	634	308.5	708.6	1651
Germany	71.6	111.5	183.2	356.7	197.1	352.9	906.6
Greece	0	58.5	58.5	71.9	27.3	105.4	204.6
Ireland	0	6.8	6.8	17.5	9.3	122.1	149
Italy	13.4	135.1	148.5	183.5	109.5	293.8	586.8
Netherlands	23.3	15.8	39.1	15.6	8.4	58	82
Portugal	5.8	93.3	99.1	131.4	7.7	26.6	165.7
Spain	21.6	140.7	162.3	265.4	96	294.8	656.1
Sweden	65.3	209.5	274.8	655.9	29.9	59.1	744.9
United Kingdom	26.5	12.6	39.1	70.7	108	397.5	576.2
Total	370.3	1454.8	1825.2	3144.8	1001.1	2613.4	6759.2

Source: Kaltschmitt and Bauen (1999).

solid biofuels. This includes the agricultural production of biofuels, their preparation and provision, their use in the generation of energy and in the recycling of ashes. Additionally, noneconomical and nontechnical barriers that slow down a widespread use of solid biofuels need to be addressed.

Compared to other renewable energy sources, solid biofuels are characterized by a wide range of fuel types. They differ in origin, physical and mechanical properties (e.g. moisture content, particle size and particle size distribution) and chemical composition (e.g. content of sulfur, nitrogen and chlorine). In fact, lack of clearly defined biofuel properties as well as clear supply conditions are seen as major nontechnical barriers for biofuels (Kaltschmitt et al., 2001).

Thus standardization of biofuel properties and their measurement is one of the tools that needs to be developed to improve biofuel markets. Standardization is expected to improve markets in the following ways:

- Producers of solid biofuels get more concrete instructions for the production of solid biofuels. They are then able to optimize their production processes with regard to the properties demanded of the fuels and can reduce costs through a more efficient production.
- Having a solid specification available, one that is well adapted to practical needs, the development of a solid biofuel market is more promising. The

properties of the trade product *solid biofuel* are clearly defined and well known just as it is, for example, for different liquid fuels such as gasoline or fuel oil. Prices will then reflect specific categories and qualities of the fuels, these also being well defined and well known. This makes the markets more transparent, favoring cost reduction and volume increase.

- Energy provision systems and conversion technologies can be better designed and optimized to operate more efficiently and environment-friendly if fuel quality is defined within a narrow range. This refers primarily to the requirements concerning conveyor problems, emissions control or corrosion phenomena.

There is a general agreement on the need for European standards in the field of solid biofuels. European standards are seen as a good tool to develop business opportunities and acceptance in the area of biomass. In particular, countries with a high potential share of solid biofuels regard standardization as an important step in promoting the use of biomass as energy source.

11.2. WHAT SHOULD BE STANDARDIZED?

National standards, best practice lists or quality assurance manuals in the field of solid biofuels have already been developed and implemented in different countries within the European Union (e.g. Austria, Finland, Germany, the Netherlands, Sweden). This has contributed to improve the matching of fuel quality and fuel provision with conversion equipment and systems, facilitate the comparison of the quality and value of different solid biofuels, and determine and assure fuel quality.

A review of the existing national standards, best practice lists or quality assurance manuals allows for some observations regarding the harmonization of standards on terminology and classification that is needed at the European level. Table 11.2 gives an overview of the various biofuel properties which are relevant for assessing biofuel quality, and which are considered within the different EU countries. First, there are fuel properties focusing on the physical and mechanical condition of the fuel and its behavior during biofuel handling (loading, transportation, storage, feeding). Second, there are fuel properties influencing the process of energy conversion (e.g. combustion or gasification). And last but not least, there are chemical properties needed to calculate the flow of molecules and to assess the emissions.

With regard to the intended effects of standardization procedures at the European level, there are two main relevant fields to be considered.

1. *Definition of sources and types of solid biofuels through a detailed and transparent terminology of the biomass resources (i.e. the different types of forest products and residues).*

Table 11.2. Fuel specifications to characterize solid biofuels

	Austria	Switzerland	Germany	Finland	Italy	Netherlands	Sweden	Total
Moisture	x	x	x	x	x	x	x	7
Particle shape/size	x	x	x	x	x	x	x	7
Heating value	x	x	x	x	x	x	x	7
Provenance	x	x	x	x	x	x		6
Ash	x		x	x	x	x	x	6
Density	x	x	x				x	4
Sulfur			x	x		x	x	4
Chlorine, Fluorine			x			x	x	3
Nitrogen			x	x		x		3
Volatiles				x		x	x	3
Durability	x						x	2
C, H, O				x		x		2
Major elements				x		x		2
Minor elements				x		x		2
Ash melting point				x				1
C/N-ratio					x			1

Source: Rösch et al. (2000).

The terminology, definitions and descriptions as well as the fuel specifications and classes which are applied in different EU countries today are the product of the traditions and characteristics, specific to the national fuel market and their information requirements. Problems of comparability of national terms and definitions can arise due to differences in the national systems of nomenclature. These nomenclature differences make fuel comparison across EU countries difficult and serve as obstacles for trade. Standards of terminology and definitions as well as fuel specifications and classes can make a valuable contribution towards allowing a more direct assessment of the quality and value of solid biofuels, thus fostering trade and a broader use of biofuels as an energy source throughout Europe. Standards can also provide a vital input to facilitate broad resource and market assessments (e.g. internationally comparable results and statistics for solid biofuels) which can help shape public and industrial policy.

2. *Identification and classification of the most relevant biofuel properties to ensure a cheap and trouble-free conversion with low emission levels.*
   The number of fuel properties used in national standards, quality guidelines and assurance manuals or recommendations to classify solid biofuels varies between 3 and 90 for wood chips. The number of classes for fuel pellets and briquettes varies between 3 and 5 only because of the compression process which makes the fuel much more homogenous in size and shape, moisture content and energy density. In Germany, for example, only a wood pellet standard is available (DIN 51731, 1996).

In EU countries, only a few important fuel properties are presently used to assess fuel quality and value (among them the moisture content, the particle size and the heating value) due to practicability and costs of performing fuel property measurements. A review of the existing national standards and practices indicates that an intended European classification system of solid biofuels should satisfy the following criteria (Rösch et al., 2000):

- The biofuel classification system should be universal and comprehensive so that it can be used for all kinds of biomass for energy generation;
- Classification should be restricted to the biofuel properties which are most relevant in practice, being clearly defined and easy to control;
- The range and amount of classes for each of the biofuel types should take into consideration regional diversity as well as the specific demands of different provision and combustion technologies and trading practices;
- Each biofuel should be classifiable by a plain and easy code (i.e. like the classification of coal, fuel oil or steel);
- Conversion plants using solid biofuels to produce green electricity should be able to use the biofuel class code for certification of their products.

11.3. BUILDING A SOLID BIOFUEL STANDARDIZATION PRACTICE IN EUROPE

The benefits of developing European standards as a means to stimulate solid biofuel utilization and trade has gradually become more apparent to many players in the bioenergy market. As a result, the European Committee for Standardisation, CEN, was given the mandate to initiate the development of European standards for solid biofuels. A work program for a CEN Technical Committee for solid biofuels was drafted and approved within a FAIR/THERMIE[1] consortia sponsored by the EC Directorate General for Research and Energy. The program was based on standardization reports in all EU countries. Finally, a Technical Committee for Solid Biofuels was established at the end of May 2000 to undertake the work.

The Technical Committee should propose standards applicable to solid biofuels that originate from the following sources (CEN TC 335, 2001):

- Products from agriculture and forestry;
- Vegetable waste from agriculture and forestry;

[1] FAIR, Agriculture and Agro-Industry including Fisheries Programme of Research and Technological Development, was implemented under the Fourth Framework Programme of the EU (1994–1998). THERMIE, the demonstration component of the nonnuclear energy RTD Programme of the EU was implemented in the period 1995–1998.

Table 11.3. Working groups of CEN TC 335 solid biofuels

No.	Working group	Convenor of working group
I	Terminology, definitions and description	Deutsches Institu für Normung e.V. (DIN), Berlin, Germany
II	Classification and quality assurance	Suomen Standardisoimisliitto r.y. (SFS), Helsinki, Finland
III	Sampling and sample reduction	British Standards Institution (BSI), London, United Kingdom
IV	Physical mechanical testing	Standardiseringen i Sverige (SIS), Stockholm, Sweden
V	Chemical testing	Nederlands Normalisatie-Instituut (NEN), Delft, The Netherlands

DIN: German Institute for Standardization.
SFS: Finnish Standards Association.
SIS: Swedish Standards Institute.
NEN: Netherlands Standardization Institute.

- Vegetable waste from food-processing industry;
- Wood waste, with the exception of wood waste which may contain halogenated organic compounds or heavy metals as a result of treatment with wood preservatives or coating (includes particularly wood waste originated from building and demolition waste);
- Cork waste.

The activities of CEN TC 335 Solid Biofuels are accompanied by the so-called Mirror Committees within the CEN member states (i.e. Germany, Austria, Sweden). Through these committees, all the concerned national key actors (i.e. manufacturers of equipment, traders, consumers, scientists) get the possibility to join the standardization process and add their experiences to the work. This is very important since the European standards, when they come into force, will overrule the existing national standards.

Five working groups were established to develop more than 20 European standards (see Table 11.3). The working groups are formed by different European experts and started their work in late 2000. The first draft standards were forwarded to the Mirror Committees by the end of 2002 for voting. The first implementation phase at the European level was expected during 2004. Table 11.3 shows the role of the working groups and who is leading each task. The tasks are further explained here.

Terminology, definitions and description

The development of a terminology standard is based on the fact that biofuels are produced from different sources, have different nature, types and properties, and that the purpose is its conversion into bioenergy. Terms for sampling, testing and classification are also important. A strong cooperation takes place with the other working groups to guarantee that the final terms chosen are the most relevant.

Classification and quality assurance

The classification system requires a detailed terminology of the biofuel sources, which considers not only the kind of biomass (woody biomass, herbaceous biomass etc.) but also its origin (i.e. logging residues, energy plantation wood, whole trees). This information mentions important conclusions on the biofuel properties (i.e. the ash content of forest wood mainly depends on the amount of bark).

However, besides the terminology, some biofuel properties have to be further classified. To identify the biofuel quality, the shape and size, density, moisture content and ash content of the parameters are reckoned as the most relevant (Rösch and Kaltschmitt, 2001; Rijpkema, 2001). These are further explained here. The importance of other biofuel properties (i.e. content of different elements, ash melting behavior) depends on the type of solid biofuel, the specific conditions at the combustion plant, the emissions control etc. For most of the currently used woodfuels, these properties have no significant relevance and thus should be taken into consideration only under particular circumstances. Approaches to quality assurance are discussed in the next section and exemplified through the case of straw.

Shape and size. The mechanical properties of solid biofuels are relevant for transportation and reaction at the conversion plant. In practice the shape and size vary widely i.e. between milled biofuels (i.e. wood flour), compressed biofuels (i.e. straw pellets), cut biofuels (i.e. chips) and baled biofuels (i.e. straw bales). Those different types of fuels need specific equipment for production, transportation, storage, feeding and combustion. For example, the trouble-free handling of chips is limited by a certain amount of over-sized particles as well as a certain amount of very fine particles (dust). A wide range of particle sizes can cause trouble in fully automated feeding systems due to bridging, obstruction or adhesion.

Density. There are two different types of density which are relevant for solid biofuels. The particle density describes the density of the material itself and is relevant for the combustion process (i.e. evaporation rate, energy density etc.), some feeding aspects (i.e. for pneumatic equipment) and storage. The particle density can only be varied by producing compressed biofuels and is used to describe the quality of those products (i.e. high particle density is an indicator for a high pellet quality). The bulk density is defined as the ratio of dry material to bulk volume and is relevant for the volume needed for transportation and storage.

Moisture content. The moisture content of solid biofuels varies within a wide range. For example, the moisture content of woodfuels depends on the time of harvesting, the location, type and duration of the storage and the fuel preparation. It varies from less than 10 per cent (residues from wood processing industry) up to 50 per cent (forest wood chips). The moisture content is relevant not only for the heating value but also for the storage conditions, the combustion temperature and the amount of exhaust gas.

Ash content. The ash content of solid biofuels depends on the type of biomass and the impurities. It is relevant for the heating value and to decide whether the biofuel is appropriate for use in particular combustion plants.

Sampling and sample reduction

The European standards for sampling, sample reduction and sample treatment are regarded as important and this area seems to be a weak point within the process to determine fuel properties. There are doubts in different EU countries whether the samples drawn from the lorry or conveyor belt can be regarded as being representative for the delivered fuel load. The moisture content, which is the main fuel property to measure the value of the fuel, can show great variation within a single truckload. If the sample shows a different moisture content from that of the average of the truckload, the calculation of the energy content and thus the calculated price to be paid for the biofuel can be misleading.

Physical–mechanical and chemical testing

For all or most of the tests to determine the physical–mechanical and chemical fuel properties, standards are important to assure the comparability of results. Most of the fuel deliveries are settled on the basis of on-site measurements to determine moisture content, particle size and size distribution and other main fuel properties. Therefore, there is a strong requisite for standards for the on-site tests. The tests should be suitable to determine the fuel properties and to assess the fuel quality and value at the point of fuel delivery. These on-site test methods do not need to have the same accuracy as the actual laboratory methods but rather be simple, yet reliable and inexpensive, while being a representative assessment of the fuel value. Smaller plant owners have been reluctant about this idea, as they fear that new elaborate standards for sampling and testing may force them to hire external consultants and services from specialized laboratories. This would obviously raise the costs for bioenergy.

11.4. QUALITY ASSURANCE – EXAMPLE OF STRAW QUALITY IMPROVEMENT

Activities on standardization have to be followed by the development of quality assurance systems relevant for biofuel provision and utilization. Currently, there is no quality assurance system that takes into account the whole provision chain of solid biofuels. A standard for quality assurance will be developed within CEN/TC335/working group II (see Table 11.3).

Theoretically, quality assurance systems can be introduced in all processes of the provision chain. Figure 11.1 gives an overview of the biofuel chain, the different processes, the basic conditions and an illustration of factors influencing quality along the chain. In practice, the first step is to identify the points where the relevant physical and/or chemical parameters can be measured and controlled easily.

A motivation for quality assurance is to guarantee that the processes follow environmental laws (e.g. emission limits) or meet technical requirements at the conversion plants (e.g. avoiding corrosion). The quality of the fuel can be controlled, for example, when chemical and/or physical parameters are modified in crop production processes (e.g. modifying the nitrogen content of whole grain crops by nitrogen fertilization) or in harvesting and fuel preparation (e.g. modifying the water content of wood by storage). At the conversion plant, technical solutions are available for emissions reduction (e.g. primary and secondary measures for NO*x*-emissions reduction).

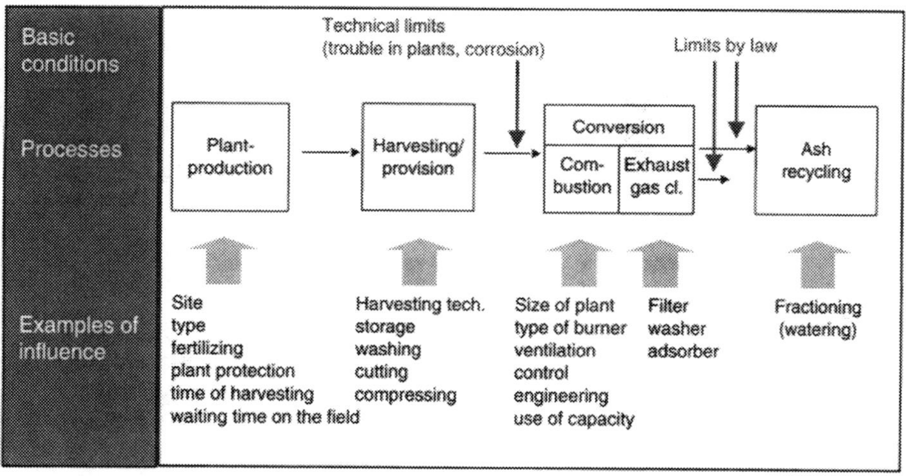

Figure 11.1. Principle of a solid biofuel chain, including basic conditions, processes and examples of factors influencing quality. Source: Thrän et al., (2001).

Measures that affect quality should be identified preferably where the costs are lower along the chain. Although different control points are practicable, it is believed that a quality assurance system shall start at the combustion plant, in cooperation between the fuel traders and combustion plant operators, which will ensure a trouble-free and low emission business. Additionally, the manufacturers of equipment for biofuel handling and combustion should take quality assurance among their considerations in the technical development of their equipment.

Specific research is being conducted to optimize the biofuel production chains and to recommend the most promising biofuel classes (respectively supply chains) for different markets. In Germany, for example, various options to modify solid biofuel characteristics such as nitrogen and water content have been identified within the production and provision of certain biofuels. This includes fertilization and storage practices. Also, options to reduce emissions within biomass burning have been studied (Thrän et al., 2001). An example for the case of straw is provided here to illustrate how quality improvements can be accomplished.

Theoretically, different options are available to modify the *chlorine* content of straw respectively, the HCl-emissions in crop production, harvesting/preparation as well as conversion. Figure 11.2 illustrates measures for reducing *chlorine* content in

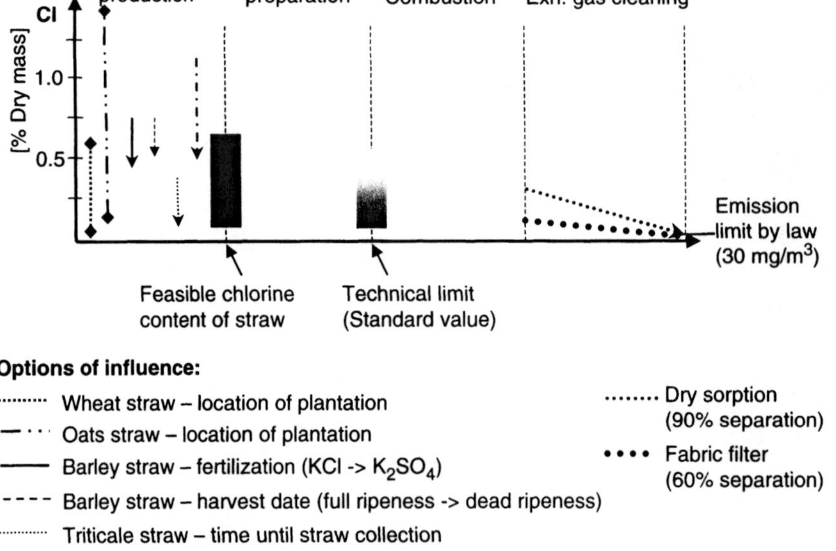

Figure 11.2. Factors influencing chlorine content in straw. Source: Obernberger (1996); Vetter and Hering (1999, 2000, 2001).

straw. It follows the different steps of information along the chain of production, provision and energetic utilization of straw. The arrows give an overview of straw chlorine and HCl reduction measures potentially available, especially within crop production and biofuel conversion. The first hatched rectangle describes the feasible chlorine content that can be achieved through crop production measures (0.1 to 0.6 per cent). The second rectangle indicates the maximum tolerable chlorine contents (0.1 to about 0.4–0.5 per cent), which are derived from technical limits set by plant corrosion problems and/or emissions regulations. These measures for *chlorine* reduction are discussed in turn.

Crop production. One way to modify *chlorine* content is to make a targeted selection of straw for energetic purposes. Investigations show that the *chlorine* content in wheat straw varies within a lower range than oats straw. Fertilization of the grain has an important influence on straw biofuel quality. A K2SO4-fertilizer, for instance, reduces the *chlorine* content of straw by 0.3 per cent as compared to the application of a KCl-fertilizer (Vetter and Hering, 1999).

The harvest date also influences the *chlorine* content. Harvesting the grain at the time of dead ripeness[2], compared to the time of full ripeness, allows further *chlorine* reduction (Vetter and Hering, 2000). Another option is to extend the period between harvest and straw collection from the field. One reason why this can reduce *chlorine* is that the straw may be washed by rain. Vetter and Hering (1999) observed that the *chlorine* content from (winter) barley straw was reduced from 1.12 to 0.5 per cent based on a three week "storage" on the field before straw collection (baling). Comparable conditions provided a *chlorine* content reduction from 0.4 to about 0.13 per cent for triticale straw.

Preparation and harvesting. Within the process of straw preparation, the *chlorine* content may be washed out by technical means (Nikolaisen et al., 1998). Investment costs are on the order of DKK 200 million (Danish crowns)[3] for a plant that "cleans" about 125 000 to 150 000 tons of straw per year. Such high costs indicate that this measure is not very attractive as a first step, the reason why it is not considered in Figure 11.2.

Conversion. Filters are often required at the conversion plant to reduce dust emissions from the stack. At the same time, these filters contribute to reduce HCl emissions. Another measure for HCl reduction is the use of a sorbent (e.g. lime

[2] This is the period after full ripeness and the best period for harvesting corn. At this point, the water content of corn is about 14–16 per cent and the straw becomes fragile.

[3] Approximately 26.9 million Euro.

hydrate). This allows for about 90 per cent reduction of HCl emissions, as the *chlorine* is incorporated in the ash.

The example of quality assurance of the *chlorine* content in straw shows that there are many options to accomplish improvements in the crop production field. A lower *chlorine* content reduces technical risks such as corrosion at the plant, as well as emissions. Thus a quality assurance system for straw has to include crop production measures besides the options available at the conversion plant.

However, a clear and detailed description of the various steps and aspects of a quality system has to take into account the local and regional conditions for crop production, the economic implications of the various measures, and the practical experiences of the various stakeholders (farmers, supervisor at the conversion plant). More knowledge on economic and technical advantages of the different measures and processes shall allow for a more specific discussion on the options available at the regional and national levels.

11.5. FINAL REMARKS

There is a significant potential all over the EU for an increased use of biomass. There is a general agreement in Europe upon the need to develop European standards for solid biofuels as a way to foster the utilization of the existing potential and further development of biofuel markets. Therefore, standardization of biofuel properties and definition of methods for their measurement is underway. More than 20 standards are being considered and specific attention is being given to terminology, classification and quality assurance. While the discussion on terminology and classification has moved fast towards national discussions for adjustments, the development of quality assurance will require additional research. Investigating the biofuel chain at the national level is one systematic approach to find out more about different options for an efficient and global quality assurance taking into account the specific national conditions.

REFERENCES

CEN TC 335 Solid Biofuels (2001) Work Programme for solid biofuels and draft Business Plan, available in the internet at www.bio-energie.de, link Feste biomasse/Standardisierung, on May 29, 2005.

Country reports from Austria, Belgium, Denmark, Finland, France, Germany, Greece, Great Britain, Ireland, Italy, Netherlands, Sweden, Spain, available in the internet at www.bio-energie.de, link Feste biomasse/Standardisierung, on May 29, 2005.

DIN 51731 (1996) Presslinge aus naturbelassenem Holz, Anforderungen und Prüfung (Compressed Untreated Wood, Requirements and Testing), Beuth Verlag, Berlin.

Kaltschmitt, M. & Bauen, A. (1999) Contribution of biomass towards CO_2 reduction in Europe (EU), *Proceedings 4th Biomass Conference of the Americas*, Oakland, California, USA.

Kaltschmitt, M., Rösch, C. & Thrän, D. (2001) Normung biogener Festbrennstoffe – Notwendigkeiten und laufende Normungsaktivitäten (Standardisation of Solid Biofuels – Requirements and On-going Standardisation Activities), in Ed. Fachagentur Nachwachsende Rohstoffe (FNR), Eigenschaften biogener Festbrennstoffe (Solid Biofuel Properties). Schriftenreihe Nachwachsende Rohstoffe, Band 17, Landwirtschaftsverlag Münster, pp 5–37.

Nikolaisen, L., Nielsen, C., Larsen, M.G., Nielsen, V., Zielke, U., Kristensen, J.K. & Holm-Christensen, B. (1998) *Straw for Energy Production*, 2nd Edition, Technology – Environment – Economy, The Centre for Biomass Technology, Denmark.

Obernberger, I. (1996) Erforderliche Brennstoffeigenschaften holz- und halmgutartiger Biomasse für den Einsatz in Großanlagen (1 bis 20 MW thermischer Leistung) (Required properties of woody and herbaceous biomass for use in plants with thermal output of 1 to 20 MW), in Ed. Fachagentur Nachwachsende Rohstoffe (FNR), Biomasse als Festbrennstoff (Biomass as a Solid Biofuel), Schriftenreihe Nachwachsende Rohstoffe, Band 6, Landwirtschaftsverlag Münster, pp 122–154.

Rijpkema, B. (2001) Classification system for biomass fuels – The Dutch Example, in Ed. Fachagentur Nachwachsende Rohstoffe (FNR), Eigenschaften biogener Festbrennstoffe (Solid biofuel properties), Schriftenreihe Nachwachsende Rohstoffe, Band 17, Landwirtschaftsverlag Münster, pp 285–295.

Rösch, C., Kaltschmitt, M. & Limbrick, A. (2000) Standardisation of solid biofuels in Europe, *Proceedings 11th European Biomass Conference and Exhibition*, Sevillia, Spain.

Rösch, C. & Kaltschmitt, M. (2001) Normen für biogene Festbrennstoffe – Analyse vorhandener Ansätze und Vorschlag für ein Klassifikationssystem für Europa (Standards for solid biofuels – analysis of existing standards and proposal of a European classification system), in Ed. Fachagentur Nachwachsende Rohstoffe (FNR), Eigenschaften biogener Festbrennstoffe (Solid biofuel properties), Schriftenreihe Nachwachsende Rohstoffe, Band 17, Landwirtschaftsverlag Münster, pp 296–311.

Thrän, D., Härdtlein, M., Kaltschmitt, M. & Lewandowski, I. (2001) Ansatzstellen für die Normung biogener Festbrennstoffe – Identifikation und Quantifizierung geeigneter Kenngrößen und Schnittstellen (Starting points for standardisation of solid biofuels – identification and quantification of practicable fuel properties), in Ed. Fachagentur Nachwachsende Rohstoffe (FNR), Eigenschaften biogener Festbrennstoffe (Solid biofuel properties), Schriftenreihe Nachwachsende Rohstoffe, Band 17, Landwirtschaftsverlag Münster, pp 260–284.

Vetter, A. & Hering, Th. (1999) Beitrag der Thüringer Landesanstalt für Landwirtschaft im Rahmen des 2. Zwischenberichts zum BMVEL/FNR Projekt Voraussetzungen zur Standardisierung biogener Festbrennstoffe (Az.: 97NR055) (Contribution of the Thüringer Landesanstalt für Landwirtschaft within the Second Interim Report to the BMVEL/FNR Project Preconditions to Solid Biofuel Standardisation) (not published).

Vetter, A. & Hering, Th. (2000) Beitrag der Thüringer Landesanstalt für Landwirtschaft im Rahmen des 3. Zwischenberichts zum BMVEL/FNR Projekt Voraussetzungen zur Standardisierung biogener Festbrennstoffe (Az.: 97NR055), (Contribution of the Thüringer Landesanstalt für Landwirtschaft Within the Third Interim Report to the BMVEL/FNR project, preconditions to Solid Biofuel Standardisation), Interim report (not published).

Vetter, A. & Hering, Th. (2001) Einfluss von Art, Standort und pflanzenbaulichen Maßnahmen auf die Eigenschaften annueller Pflanzen (Influence of species, location and of agricultural measures on annual crops properties), in Ed. Fachagentur Nachwachsende Rohstoffe (FNR), Eigenschaften biogener Festbrennstoffe (Solid biofuel properties), Schriftenreihe Nachwachsende Rohstoffe, Band 17, Landwirtschaftsverlag Münster, pp 118–133.

Part IV

Exploring Opportunities through the Clean Development Mechanism

Chapter 12
The Clean Development Mechanism (CDM)

Semida Silveira

12.1. THE CHALLENGE OF MITIGATING CLIMATE CHANGE

The climate change problem results from the concentration of greenhouse gases in the atmosphere, mainly an effect of industrial development and fossil fuel utilization over the last two centuries. The initial milestones in addressing climate change were the signature of the Climate Convention (UNFCCC) in 1992, and the negotiation of the Kyoto Protocol in 1997. The overarching objective of the Climate Convention is the stabilization of greenhouse gas emissions at levels that can prevent dangerous human interference with the climate system (UNFCCC, 1992). The Kyoto Protocol defines steps in the implementation of the Climate Convention (Kyoto Protocol, 1997). Besides defining the emissions reduction target of 5 per cent compared with 1990 levels, the Kyoto Protocol sets the institutional basis for the formation of greenhouse gas markets, and creates mechanisms for the full participation of all parties to the Convention and the Protocol in the implementation of agreed objectives.

Today, many governments consider the mitigation of climate change the most difficult challenge of the next few decades. One of the underlying difficulties is the broad nature of the problem which refers not only to the globality of the natural phenomena that affects climate on earth, but also to the implications that climate change mitigation and adaptation may have on the international economy and development in general. Another difficulty is that the issue of environmental sustainability, climate change included, requires the scrutiny of development strategies, technologies, consumption behavior, life styles, and even the institutions that have served as the basis to build modern society. Thus the climate change problem requires global solutions, and efforts from all nations.

Measures to mitigate climate change touch the basis of economic development, as mitigation measures are likely to have impact on the way major sectors such as energy, transportation, agriculture and forestry operate and are further developed. It is difficult to calculate with precision the overall costs of mitigating climate change, although many efforts have been made in this direction. Initially, mitigation measures were perceived as very costly and extremely risky for the global economy

Bioenergy – Realizing the Potential

due to the number of interventions and coordination of efforts necessary. This made many corporations, governments and specialists extremely negative about the whole issue. However, after a decade of debates and initial attempts to implement mitigation measures, a consensus has evolved around the need to incorporate climate change issues into public and private development strategies.

This general consensus has served not least to engage the private sector into the climate change debate in a fruitful way. Yet, many efforts are still needed to make a quantitative change in the current scenario of greenhouse gas emissions. In a world of unequal development, a major challenge is to find a fair way of distributing the burden of measures among nations. How to shift the development path so as to achieve a scenario of emissions stabilization without constraining the development of poor nations? The projected growth of developing countries implies significant increases in greenhouse gas emissions which will have to be partly mitigated through the choice of low-emissions technology systems and partly compensated with reduced emissions in industrialized nations. How to accomplish that?

In an initial effort, the Kyoto Protocol attempts to create an institutional platform for cooperation among nations in measures to mitigate climate change. Three mechanisms, the so-called flexible mechanisms, are created: emissions trading (ET), joint implementation (JI), and clean development mechanism (CDM). The operation of these three mechanisms aims at reducing mitigation costs and accelerating the pace of emissions reduction. Emissions trading allows for carbon credits to be traded among companies and countries to facilitate meeting emissions reduction targets. Joint implementation provides the basis for industrialized countries and economies in transition to collaborate in achieving targets jointly. Finally, the CDM provides the opportunity for developing countries to participate in projects to reduce actual or expected greenhouse gas emissions.

The work on flexible mechanisms has engaged many academics and specialists intensively, and the literature on the subject is already vast. We make no attempts to review all the issues involved in the climate change debate, or to analyze possible outcomes of the flexible mechanisms. This chapter provides only a brief introduction to the context of the global climate change agenda and to CDM in particular. Together with Chapter 13, we hope that it will help readers appreciate the examples of CDM projects provided in Chapters 14, 15 and 16. The latter are pioneer projects developed in Brazil and Ghana which give an idea of how CDM can contribute to promote bioenergy in developing countries.

12.2. THE CONCEPT OF CDM

The Clean Development Mechanism (CDM) creates an institutional base for a direct participation of developing countries in the implementation of the Kyoto

Protocol. Although no emissions reductions target has been negotiated to developing countries, their involvement in climate change activities is essential since these countries are expected to have a huge increase in greenhouse gas emissions in the coming decades. Through the CDM, developing countries can become fully involved in mitigation measures, and thereby, important actors in the formation of carbon markets. This shall help pave the way to more ambitious targets and new commitments at a later stage.

The two overall requirements of CDM projects is that they should contribute both to the reduction of emissions according to a baseline or predetermined scenario, and to sustainable development according to priorities and strategies defined by the host country. The baseline gives the trajectory of expected emissions in the absence of the project. The emissions reductions have to be quantifiable and measurable, and should result in improvements in relation to the baseline. The contribution to sustainable development and to reduce emissions entails the issuing of certificates, CERs, which can be traded internationally after proper validation of the project, and verification of the emissions reductions actually achieved. Figure 12.1 summarizes the key requirements of CDM projects.

Various CDM projects are being presently implemented around the developing world. This is contributing to the development of methodologies to define baselines and determine actual emissions reduction. While baselines seem to be one of the most difficult issues around the CDM, once determined, the measurement of emissions reductions is a rather technical matter. It is less clear, however, how the sustainability dimension associated with these projects will be more closely defined and monitored. In principle, the contribution of the project to sustainable development will be measured in relation to the host country's development priorities and strategies. On the other hand, the long period for CER accreditation in a CDM project may be in contrast with the nonlinear dynamics of socioeconomic development within the same time frame, and the still very recent experience of introducing sustainability criteria in development strategies and projects.

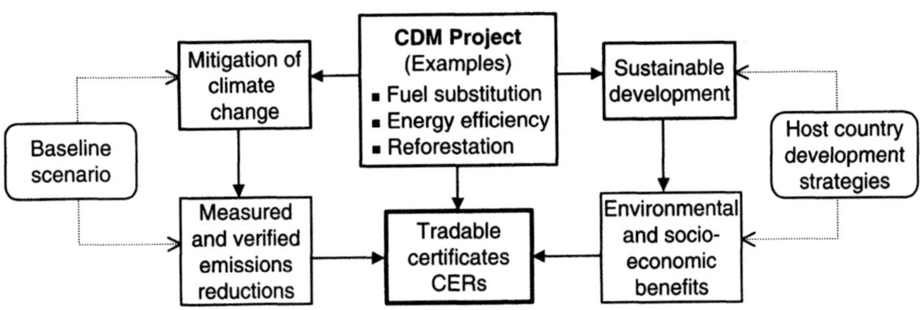

Figure 12.1. Key requirements of CDM projects.

12.3. THE CDM PROJECT CYCLE AND INSTITUTIONAL FRAMEWORK

The CDM project can be developed by public or private entities. The objective is to create a clear line of procedures so that participation is facilitated, and CERs can become internationally accepted and tradeable. Initially, CDM projects were mainly developed on an experimental basis by enthusiastic countries and companies that believed in the economic and environmental benefits of CDM, or wanted to take advantage of being early starters. Attitudes towards the mechanism have varied both in developing and industrialized countries, but ratification of the Kyoto Protocol has contributed to broader engagement of governments and enterprises in developing new projects.

A project becomes a CDM project when conceptualized according to the requirements of the mechanism. A project design document (PDD) is prepared including the description of project activities and participants, project boundaries, the baseline and methodology for quantifying the reduction of emissions, expected leakages, as well as a plan for monitoring and verifying those reductions. The additionality of the project needs to be justified. This means that the project should result in emissions reductions which would not occur in the absence of the project.

The CDM project needs the approval of the host country's CDM authority, the so-called Designated National Authority (DNA), which shall certify that the project contributes towards sustainable development. After this, the project needs to be registered with the Executive Board in order to be eligible for CERs. The Executive Board supervises the implementation of the CDM and is nominated by the parties to the Kyoto Protocol.

The project is implemented and monitored by the developers and other participants, who shall collect all necessary information for the verification of emissions reduction by authorized entities. The Designated Operational Entities are the only ones in position to validate the project activities and performance, and certify the accomplishment of emissions reductions. Only then can CERs be issued by the Executive Board and, eventually, traded or used to meet commitments. Figure 12.2 summarizes the CDM project cycle. The upper boxes indicate the institutions responsible for each step.

The role and level of importance attributed to CDM as part of emissions reduction efforts has varied among industrialized countries. A country can choose to use CERs as part of mitigation measures of companies in their territory, or can put a cap to how much reduction can be accomplished through CDM projects. Ideally, there should be as many projects as possible to reduce emissions as much as possible and foster sustainable development. In practice, however, industrialized countries need to also implement measures at home to enhance credibility about their

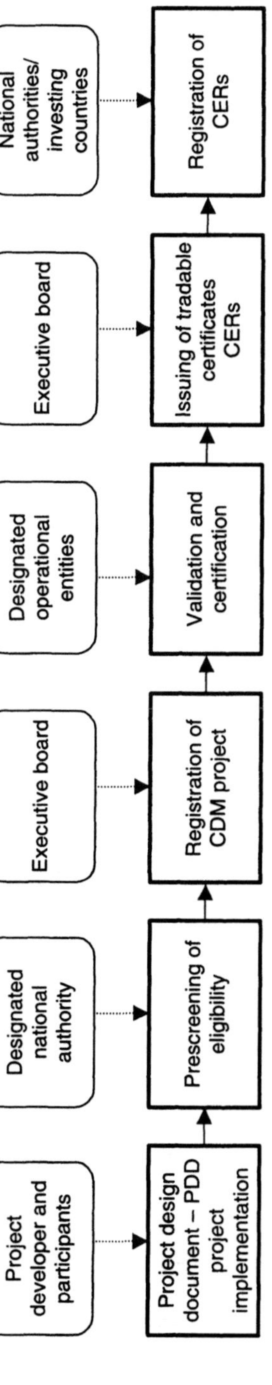

Figure 12.2. CDM project cycle and responsible authorities.

commitment to climate change issues, and as a way to foster low carbon or carbon-free production systems.

12.4. WHO WILL PARTICIPATE IN CDM PROJECTS AND WHY?

We have previously pointed out the need to involve developing countries more directly in climate change mitigation measures. Energy demand in developing countries is expected to at least double in the next 20 years which shall result in significant increase in greenhouse gas emissions (Nakicenovick et al., 1998). The dynamics of well-established industries and fuel markets shall strongly influence the evolution of energy generation and its use in developing countries in the medium term, leading to an overall increase in the utilization of fossil fuels and, consequently, more emissions from these countries.

However, there are opportunities for developing countries to take a different direction. Already today, medium-income countries are moving towards larger energy efficiency and increased utilization of combustible renewables and waste (Sun, 2003; Miketa, 2001). Liberalization of energy markets and various national policies have already led to significant savings of emissions in fast-growing developing countries such as Brazil, India and China. These countries already constitute important markets for clean technologies. Therefore, it is important that, as new investments are made in the expansion of energy supply infrastructure in developing countries, renewable systems be given high priority and sustainability criteria be observed. In this context, CDM can contribute as a channel to attract capital aimed at clean technologies that contribute to socioeconomic development, thus meeting both global environmental interests and development priority needs.

The CDM can provide a bridge for increased collaboration between industrialized and developing countries to shift energy systems towards sustainable and renewable systems. The mechanism can be used to promote renewable energy technologies and energy efficiency, reducing costs and risks and channeling investments to developing countries. To make that possible, public and private efforts have to be made jointly. The support of civil society at large is, obviously, also a prerequisite for succeeding. Therefore, stakeholder dialogs have to take place along the whole project cycle (Baumert and Petkova, 2000).

Table 12.1 summarizes the main advantages and potential barriers that different actors may find when considering involvement in CDM projects. Why should they participate in CDM? What may cause them to refrain from doing it? We have defined the actors in broad groups, differentiating among governments in industrialized and developing countries because of their different position in terms of commitments within the Kyoto Protocol. We have also differentiated companies by size,

Table 12.1. How attractive can CDM be to key actors?

	Main advantages	Barriers
Governments in industrialized countries	• Lower costs to meet commitment on emissions reduction • Help promotion of industrial development • Enhance sustainable development assistance with private sector involvement	• Development of methodologies and procedures • Maintain credibility of international community that measures are also being taken at home
Governments in developing countries	• Attraction of new capital for investment • Possibility to attract new technologies • Support promotion of sustainable development	• Mobilization of scarce resources to develop climate-related matters and CDM • Development of internal institutional capacity and procedures • Inform and provide support to local industries
Financial organizations	• Increased opportunity to finance sustainable projects with high profile • Generation of new product in the form of tradeable certificates • Lower capital risks of projects	• Long project cycle • Uncertainty of returns • Unclear rules and regulations for more rapid establishment of markets
Large corporations	• Reduction of emissions can be achieved with internal program at lower cost • Overall ecoefficiency improvement within the firm • Profile of corporate social responsibility embedded in emissions reduction measures	• Shareholder interests and pressure for short term returns • Complexity of procedures and methodologies • Transparency required of CDM project simplying less investment confidentiality
Small and medium enterprises	• Lower capital costs for projects • Possible channel to open new markets • Increased opportunity for niche markets e.g. clean technologies/opportunity to participate in package solutions in high profile international projects	• Cost of CDM component of projects • International uncertainty about the CDM regime • Lack of managerial capacity for international collaboration
International NGOs	• Transparency of methodologies and procedures for approval of emissions reduction projects • New channel to promote sustainable development	• Difficulty to develop simple rules of procedure and follow up on a variety of sectors, technologies and issues relevant to CDM
National NGOs	• Opportunity to strengthen work through link with international organizations • Attraction of new resources to work with sustainable development at local level	• Complexity of procedures and methodologies • Lack of managerial capacity • Difficulty to follow debate, acquire capacity, and develop project opportunities

separating corporations from small- and medium-sized enterprises (SMEs). SMEs can be quite different in nature depending on size, sector, type of product or service provided and even geographic location so that the advantages and barriers indicated for this group should be seen as a very broad generalization.

When it comes to financial organizations, Chapter 13 presents particular insights that complement the list in Table 12.1. We have made a distinction among international and national NGOs because of their different role in the context of CDM projects. International NGOs may have a particular impact on the development of methodologies and procedures for CDM implementation, making sure that the mechanism evolves into an effective way to facilitate the reduction of greenhouse gas emissions and to promote sustainable development. National NGOs play a role in following up the work at the local and national levels and making sure that CDM projects are really in line with the host country's sustainable development strategies. A stronger link between national and international organizations is desirable to enhance capacity building and favor resource allocation to ensure a monitoring role at national and international levels throughout the implementation of projects.

12.5. CDM AND BIOENERGY OPTIONS

Developing countries have been reluctant to accept new commitments in the implementation of the global climate agenda. This reluctance can be understood in the historical context of unequal development, national development priorities, capital shortages for new investments, and imbalances in terms of knowledge and national capacities to deal with the problem. CDM opens a window of opportunity for a stronger participation of developing countries in climate-related projects while also observing their immediate need to pursue development.

Biomass can deliver all major forms of energy, carbon is neutral if utilized on a sustainable basis, can provide a carbon sink, and contribute to large socioeconomic benefits. This makes bioenergy projects strong candidates for CDM projects. On the other hand, the need for a systems view in bioenergy solutions contrasts with the project focus of CDM. Unless well-designed development strategies and a strong multisector policy framework are provided, CDM bioenergy projects will not be able to contribute to a systems solution, or to sustainable development.

The bottlenecks for infrastructure project implementation in developing countries, including energy, are many and cannot be removed by CDM alone. Not least, the managerial and logistic requirements of bioenergy systems require moving from the technological approach often emphasized in the context of technology transfer to a managerial approach that searches for models to develop local knowledge and skills around bioenergy solutions. Again, a strong policy framework is essential to achieve such goals.

Since developing countries are very different in terms of economic development and institutional capacity, a differentiation of strategy for collaboration in CDM projects is justified (see Figure 12.3). While energy and climate policies are closely related, the focus in very poor countries should be on the provision of energy. In middle-income countries with emerging economies, the climate policy should be more strongly emphasized given that greenhouse gas emissions are increasing very rapidly in these countries. The strategy with poor countries should be to use the mechanism to form partnerships with the private sector and build technological and managerial capacity, thus enhancing development assistance programs. In emerging economies, CDM should be viewed in synergy with export policies aimed at the formation of new markets and technology dissemination, based on a policy framework to stimulate private investments in CDM projects (see also Silveira, 2005).

The CDM may help open an investment channel to develop bioenergy projects in developing countries, thereby providing an additional tool to foster wider accessibility to modern energy services in these countries utilizing indigenous energy sources. But the development of bioenergy systems in developing countries can also be considered in a broader context where developing countries become important producers of biomass to feed global systems, for example, in the production of forests and ethanol. Such an approach would require the recognition of the potential of developing countries as biomass producers, the opportunity to improve

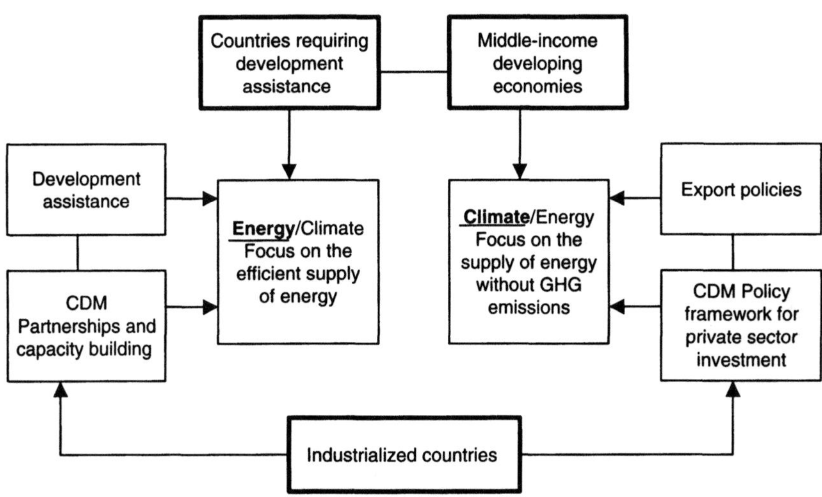

Figure 12.3. Suggested differentiated focus and synergies of energy-related CDM projects in poor and middle-income developing countries.

energy supply security at a global level, and the key role that bioenergy global solutions may play in mitigating climate change.

REFERENCES

Baumert, K. A. & Petkova, E. (2000) *How will the Clean Development Mechanism ensure transparency, public engagement, and accountability?* WRI, Washington, DC.

Clean Development Mechanism – CDM, Official information about the CDM can be found at http://cdm.unfccc.int.

Clean Development Mechanism – A Brazilian Implementation Guide, available at http://www.mct.gov.br/clima/ingles/quioto/pdf/guiamdl_i.pdf on December 08, 2004.

Danish Energy Authority (2002) Joint Implementation and Clean Development Mechanism Projects – Manual for Project Developers, Copenhagen, Denmark.

Kartha, S. & Leach, G. (2001) Using Modern Bioenergy to Reduce Rural Poverty, Report prepared to the Shell Foundation, Boston, USA, SEI.

Kyoto Protocol to the United Nations Framework Convention on Climate Change (1997) Full text available at http://unfccc.int/resource/docs/convkp/kpeng.pdf.

Miketa, A. (2001) Analysis of energy intensity developments in manufacturing sectors in industrialized and developing countries in *Energy Policy*, **29**(10), Elsevier, pp 769–775.

Nakicenovic, N. et al. (1998) *Global Energy Perspectives*, Cambridge University Press, Cambridge.

Silveira, S. (2005) Promoting bioenergy through the Clean Development Mechanism, Proceedings of the joint IEA bioenergy task 30 and task 31 workshop sustainable bioenergy production systems: environmental, operational and social implications, Oct 29–Nov 1 2002, Belo Horizonte, Brazil in *Biomass & Bioenergy*, **Vol. 28**(2), Elsevier, pp 107–117.

Sun, J. W. (2003) Three types of decline in energy intensity – an explanation for the decline of energy intensity in some developing countries in *Energy Policy*, **31**(6), Elsevier, pp 519–526.

UNFCCC – United Nations Framework Convention on Climate Change (1992) Full text available at http://unfccc.int/resource/docs/convkp/conveng.pdf.

UNFCCC (2004) A Guide to the Climate Change Convention and its Kyoto Protocol, available at http://unfccc.int/resource/guideconvkp-p.pdf on December 08, 2004.

WBCSD – World Business Council for Sustainable Development (2000) Clean Development Mechanism – Exploring for Solutions Through Learning-by-Doing, Geneva, Switzerland.

World Energy Council (2001) *Living in One World – Sustainability from an Energy Perspective*, London, Great Britain, Full text available at http://www.worldenergy.org/wec-geis/publications/reports/liow/foreword/foreword.asp on December 08, 2004.

Chapter 13

The Role of Carbon Finance in Project Development

Alexandre Kossoy[1]

13.1. INTRODUCTION

The Carbon Finance Business, CFB, at the World Bank provides a means of leveraging new private and public investment into projects that reduce greenhouse gas emissions, thereby mitigating climate change and promoting sustainable development. Carbon finance is the general term applied to financing seeking to purchase greenhouse gas emission reductions to offset emissions in the OECD countries.

Commitments of carbon finance for the purchase of carbon have grown rapidly since the first carbon purchases began less than seven years ago. The global market for greenhouse gas emission reductions through project-based transactions has been estimated at a cumulative 300 million tons of carbon dioxide equivalent since its inception in 1996 and until mid-2004. Asia now represents half of the supply of project-based emission reductions, with Latin America coming second at 27 per cent. Volumes are expected to continue growing as countries that have already ratified the Kyoto Protocol work to meet their commitments, and as national and regional markets for emission reductions are put into place, notably in Canada and the European Union.

The CFB uses money contributed by governments and companies in OECD countries to purchase project-based greenhouse gas emission reductions in developing countries and countries with economies in transition. The emission reductions are purchased through one of the CFB's carbon funds on behalf of the contributor, and within the framework of the Kyoto Protocol's Clean Development Mechanism or Joint Implementation (see also Silveira, Chapter 12). By early 2005, the CFB could count on more than US$ 850 million in nine carbon funds.

Unlike other World Bank development products, the CFB does not lend or grant resources to projects, but rather contracts to purchase emission reductions in the form of a commercial transaction, paying for them annually once they have been

[1] The views expressed in this chapter are those of the author and do not necessarily represent the views of the World Bank.

Bioenergy – Realizing the Potential

verified by a third party auditor, and the verification report delivered to the World Bank. One of the roles of the Bank's CFB is to catalyze a global carbon market that reduces transaction costs, supports sustainable development and reaches the poorer communities of the developing world.

The Bank's carbon finance operations have demonstrated opportunities for collaborating across sectors, and have served as a catalyst to bring climate issues to bear in projects relating to bioenergy, rural electrification, renewable energy, urban infrastructure, forestry, and water resource management. A vital element of this work has been to ensure that developing countries and economies in transition are key players in the emerging carbon market for greenhouse gas emission reductions. This chapter discusses how the World Bank Carbon Finance Business has dealt with project constraints, also providing concrete examples.

13.2. RISK ANALYSIS VERSUS PRICING

Project sponsors face many uncertainties before deciding to invest their time and resources in new projects. Uncertainties such as government taxation, sales quotas, limited access to new technologies, political and economic instability, subsidies available, and local currency fluctuation are among the many variables that need to be assessed. Assessing these factors becomes even more critical when considering investments in developing countries, where the impact of external factors may threaten the continuity of the business and influence the viability and success of the project.

Under different perspectives and not necessarily at equivalent proportions, lenders and borrowers share the risks involved in a financial transaction from the lender's disbursement up to the liquidation by the borrower or guarantor. At unexpected local or international financial crises during the lifetime of a loan, additional risks are added to the basket of existing risks being shared by those institutions. The likelihood of any factor to occur that might negatively affect the borrower's capacity to repay its loan is taken into account and "priced" into the total premium charged by the lender. These factors are commonly combined into the so-called Country Risk, also called sovereign and political risk.

Figure 13.1 shows a breakdown of risks, as adopted by financial institutions. The country risk includes every potential constraint for local currency convertibility to hard currency equivalents, cash transferability, asset expropriation, confiscation or nationalization of goods, governmental caps on exports (i.e. increase in local market supply), and a sudden increase in taxation on trade or cash payments abroad. Those risks are beyond the borrower's responsibility, but they largely affect their

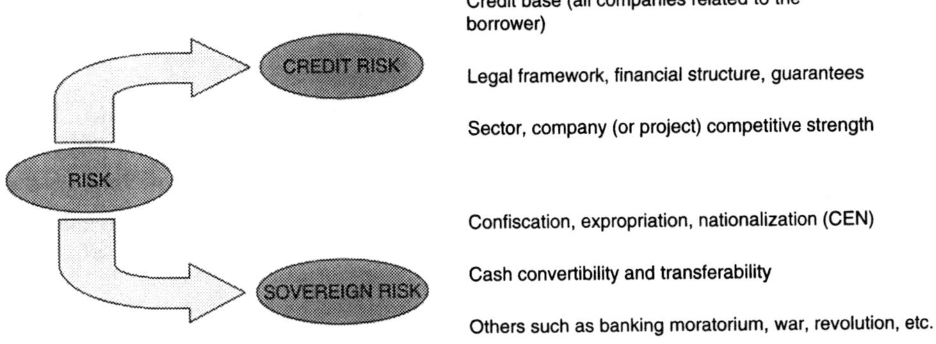

Figure 13.1. Risks related to project finance in a developing country.

capacity to produce and sell the goods being used to repay loans or they restrict the cash transfer to the lender's account.

Due to the risks involved, financial institutions set up criteria for loans in countries where those risks are more likely to happen. These criteria are normally defined in terms of a maximum cash amount available for loans. Since the risk is directly linked to the duration of the loan, more restrictive limitations are imposed on long-term transactions, unless the country risks can be mitigated. The most common way to mitigate Country Risk is by the acquisition of *Country Risk Insurance* for long-term deals from insurers, development banks or export credit agencies.

A *Country Risk Insurance* typically covers expropriation acts (confiscation, nationalization, requisition and sequestration), restrictions for currency convertibility and transfer, political violence, civil commotion, civil war, rebellion, riot, sabotage, strike, war and terrorism. Even if one or some risks are irrelevant for a specific country they are offered as part of a package, which are typically not customized. The insurance premium is directly related to the features of the transaction and the risk perception in that specific country. However, during economic turmoil, the premium of the insurance increases sharply and its availability is drastically reduced.

Therefore, a bank's internal requirement for country risk coverage can sometimes become a deal-breaker for any of the involved parties, either due to a tenor limitation for the banks (i.e. the availability of long-term funding may disappear) or due to the price limitation for the borrowers or project sponsors. Invariably, the cost for such insurance is always passed on to borrowers, thus increasing the total cost of the loan.

13.3. CARBON RIGHTS, THE EMISSION REDUCTIONS PURCHASE AGREEMENT (ERPA) AND RISK

As discussed earlier, project risks and financing prices are positively correlated. Thus factors that reduce the overall project risks will automatically reduce the price for project finance. This is what carbon finance can possibly do. We now look at the main identified risk mitigants commonly attributed to carbon rights in a project. Some of these carbon right characteristics and different roles will be further analyzed in the subsequent case-study discussion.

- ERPAs are long-term contracts denominated in hard currency. They work as a natural hedge for foreign exchange risk, and reduce the lender's exposure to local currency depreciation. This specific risk mitigation is extremely relevant for projects which operate in domestic markets and do not have access to the cheaper international loans (i.e. sources of hard currency revenue streams can increase the interest of investors and banks to participate in a project).
- Lenders who finance projects based on the production and sale of the borrower's commodities are also bearing the risk of fluctuation in the price of these goods (i.e. the same amount of goods may not cover the entire loan if the price of these goods drops). Therefore, while there is no price fluctuation for the emission reductions in the ERPAs, this can assure a constant and predictable contract value potentially able to be used as guarantee or debt-service repayment to the lender.
- Lenders are extremely concerned about the creditworthiness of the borrower's clients, who will ultimately generate the revenue streams of the borrower. The ability to provide creditworthy off-takers from a project decreases therefore some of the project-related risks. In this respect, the World Bank is considered a low risk buyer by most lenders.
- Lenders may attribute a large portion of their overall risk evaluation to the borrower's local government and actions that may hinder or prevent a loan repayment in hard currency (i.e. local currency convertibility to hard currency and transfer overseas), as well as their confiscation and nationalization of goods, and expropriation of assets, which threaten the sponsor's capacity to produce and export goods. The payment for carbon rights directly into the lender's account eliminates currency risk. Also, the existence of a government letter of approval (LoA) requested at an early stage in the process minimizes the risk of subsequent interference by the borrower's government with the generation and remission of emission reductions to buyers.

13.4. EXAMPLES OF CARBON FINANCE LEVERAGING PRIVATE AND PUBLIC INVESTMENT IN PROJECTS FROM DIFFERENT SECTORS AND COUNTRIES

The Plantar project in Brazil

The Plantar project in Brazil is one of the projects from which the Prototype Carbon Fund is buying greenhouse gas emission reductions. The Plantar project consists of the substitution of coal in the pig-iron industry (see also Fujihara et al., Chapter 14). The project aims to establish Eucalyptus plantations in degraded pasture areas. After harvesting, the timber is carbonized to produce charcoal, which is subsequently mixed with mineral iron in furnaces to produce pig iron. Due to the long lead time necessary for the eucalyptus to mature it would take up to eight years before this project could generate any cash-flow income.

Without CDM, Plantar was a project with up to eight years of implementation phase before it could start generating financial returns. In addition, three more years would potentially have been required by the project to fully pay back the investment. The project finance would require the same eight years of grace period, plus three years for amortization in order to match the usual project needs. Unfortunately, there were no loans or Country Risk Insurance available in Brazil for such a long period at any price. In these circumstances, the project was unbankable.

However, the project's eligibility to the Kyoto Protocol and the World Bank's Emission Reductions Purchase Agreement (ERPA) committing to acquire the emission reductions resulting from the project provided anticipated sources of revenue streams to be used for amortization of loan's debt service, already starting in the second year. Based on this new feature, a financial loan was structured.

In the Plantar project, the nominal value of the ERPA contract between the World Bank (as trustee of the Prototype Carbon Fund or PCF) and the project sponsor (Plantar) was anticipated by a commercial lender (Rabobank Brazil) to Plantar, the latter being both recipient of the loan and seller of the emission reductions. It was structured in a way that the expected payment for the emission reductions (in this case made by the PCF) would perfectly match the loan's amortization schedule. This transaction is similar to the common "export pre-payment" structure used in the lending sector, although it could also be correctly defined as a "monetization of the ERPA receivables". Figure 13.2 illustrates the above-mentioned financing structure.

The loan was structured in a way that the World Bank would pay for the emission reductions directly into the lender's account, therefore reducing credit and currency risk in the structure. The anticipated sources of revenue streams provided by emission reductions in the project, the absence of currency convertibility and

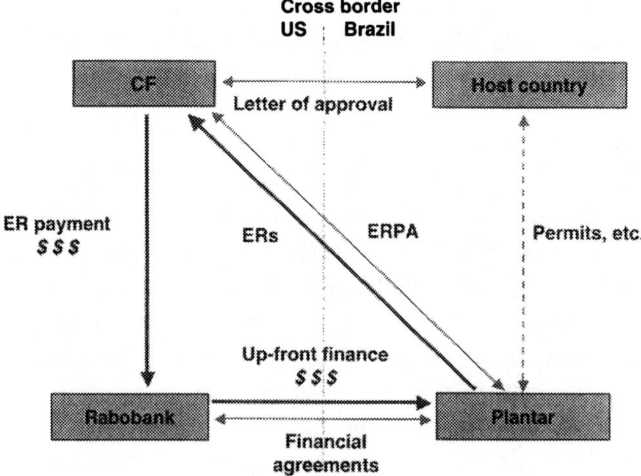

Figure 13.2. The loan structure in the Plantar deal.

transferability, and the intangibility of those emission reductions led the transaction to be rated by the lender as "Credit-risk free", resulting in the elimination of the obligation to obtain any insurance. Therefore, the project became bankable, and the loan became attractive to the lender. In addition, the credit risk mitigation also resulted in a reduction in the overall risk perception by the lender, which could provide an attractive loan to the company.

The NovaGerar landfill project in Brazil

Another carbon finance project serves as a good example of how carbon finance can play an instrumental role as a key financing tool. The NovaGerar Landfill Project consists of a sanitary landfill site being developed in southern Brazil, in which the sponsors aim to flare the methane generated on site and generate electricity from its combustion. However, as in the Plantar case, the project sponsors did not have up-front capital to invest in the required equipment.

The project sponsors could have tried to obtain a bank loan using the power purchase agreement (PPA) from the sale of energy to the grid as collateral. However, since the energy sector in Brazil has been facing serious regulatory problems since 2000, energy distributors are highly reluctant to commit themselves through long-term PPAs. Since the project's cash income was risky, its whole viability was doubtful and the project would probably have struggled to obtain financing for the necessary investment.

However, due to the emission reductions generated by the project and the World Bank's commitment to acquire all the emission reductions generated until 2012 (as trustee of the Netherlands Clean Development Mechanism Facility – NCDMF) the sponsor's supplier (i.e. a British producer and operator of flaring and energy systems) agreed to lease their equipment to the sponsor using the emission reductions income as annual payments on the lease. Therefore, the emission reductions allowed the equipment and technology supplier to provide the supplier's credit facility necessary for the project's implementation.

The high content of carbon dioxide equivalents in the methane generated by landfills resulted in an incremental project *Internal Rate of Return* of almost 25 per cent, exclusively based on the revenue streams from emission reduction. Due to the high volume of greenhouse gas emission reductions generated by the project, the carbon component not only allows for the full recovery of the supplier's investment in the flaring system, but it can also compensate potential losses in the electricity generation cash flow. The supplier agreed with the project sponsor to be paid through a percentage of the cash income from the emission reductions. The agreement between the parties has the same period as the ERPA and also requires the emission reductions payments to be made directly in the supplier's account in the United Kingdom, with a financial structure similar to that described in Figure 13.2.

The same sponsor is now being approached and has advanced negotiations with another international bank which may provide working capital resources for this project, using the revenues from the remaining emission reductions (i.e. the emission reductions not committed for the lease payment) as a loan repayment quite similar to the Plantar deal.

The Abanico Run-of-River Hydroelectric project in Ecuador

The Abanico Hydroelectric Project is a 30 MW run-of-river mini hydroelectric power plant located in Southeastern Ecuador developed by a local Ecuadorian firm. The location of the project is also one of the most economically depressed zones in the country. The project aims to generate electricity to the national grid and reliable supply of clean water to the nearby communities through a canal to be built within the project design.

Despite Ecuador's substantial hydropower capacity, which provides about 60 per cent of the country's electricity, there has been no private investment in hydropower in the country to date. This has been a function of historical risky business environment as viewed by international credit agencies, high real interest rates (14–15 per cent in US$ terms), low external capital investments, low national savings rates, and poor payment record of power off-takers (i.e. energy distributors).

In 2004, the private firm Hidrobanico S.A. developed the Abanico project. The company had assured 65 per cent of the capital expenditure through private equity from several shareholders and had sought financing from the Inter-American

Investment Corporation (IIC), the private-sector arm of the Inter-American Development Bank, for the remaining 35 per cent necessary for building the hydroelectric power plant.

Although the project had strong fundamentals such as high Internal Rate of Return (IRR) of 15.6 per cent, low investment cost of $1.1million per MW installed, capacity factor above 85 per cent, and secured Power Purchase Agreements (PPAs) with the Company's shareholders for some 35 per cent of power sales, the project fell short of IIC's investment criteria. ICC requires over 50 per cent of sales to be under firm PPA contracts and assigned to the repayment of the loan's debt service in order to mitigate credit (i.e. delivery) risks.

The project's eligibility to the Clean Development Mechanism (CDM) allowed the project sponsors to generate emission reductions along with electricity. The equity IRR increased by 0.7 per cent only with the inclusion of the emission reductions in the project's cash flow. However, the proportion of project revenues under contract reached the threshold defined by the lender, thus meeting the IIC's minimum off-take requirement.

Based on the high creditworthiness of the off-taker (World Bank), the IIC agreed to consider the proceeds of the sales of emission reductions in its investment analysis, allowing the borrower to comply with the above-mentioned covenant. According to the IIC, this played a role in securing an IIC loan and reduced the average time spent by private project developers in Ecuador to reach financial closure from the expected 5 years to less than 2 years.

Equally importantly, the financial engineering of the Emission Reduction Purchase Agreement was structured in the same way as the previous cases, so that the proceeds accrued directly to a debt reserve account in favour of IIC, thus eliminating the Ecuadorian sovereign risk. The lender's recognition of the additional benefits from this financial engineering in the loan's risk matrix was directly reflected in the loan's reduction in the interest rates by 100 base points (i.e. 1 per cent) to the borrower immediately after the ERPA signature with the NCDMF. This reduction may be translated into a cumulative economy of over US$ 300 000 in interest payments. These factors enabled IIC to extend a $ 7m, 8-year loan to Hidrobanico, facilitating financial closure for the private project.

13.5. CONCLUSIONS

In some projects in the PCF portfolio, the emission reductions are the sole source of reliable income for sponsors. It is, therefore, essential that lenders understand the value of emission reductions. This may be the trigger that will secure financing for some projects and make them viable. Special attention should be paid to the ERPA structure as explained in the text. The ERPA may and can significantly mitigate specific risks of the project, materially improving its bankability.

We looked at three different cases where carbon finance played different roles. Nevertheless, in all the three projects presented, carbon finance was of fundamental importance for the project's implementation.

- In the Plantar deal, the financial engineering in the ERPA and the anticipated revenue streams from carbon finance in the project allowed the monetization of the ERPA receivables and loan approval by a commercial bank.
- In the NovaGerar deal, the creditworthiness of the carbon credits buyer (i.e. the World Bank) allowed the project sponsors to become attractive to an international energy solution's provider, who agreed to use the carbon credits as payment for the supplier's credit offered to the project.
- In the Abanico deal, the revenue streams from carbon finance, although not relevant in terms of incremental IRR, allowed the project sponsors to reach the lender's covenant of threshold for the project's off-take contracts, resulting in the loan's approval by the lender.

In summary, the experience of these projects indicates that the qualitative value of the emission reductions in most CDM projects can exceed their quantitative value (i.e. their nominal price) as the benefits of the emission reductions and ERPAs are maximized.

Chapter 14

Cultivated Biomass for the Pig Iron Industry in Brazil

Marco Antonio Fujihara, Luiz Carlos Goulart and Geraldo Moura

14.1. THE PLANTAR PROJECT

The Plantar project has been designed according to the CDM rules agreed upon under the Kyoto Protocol. It is based on fuel switching in the iron industry, that is, avoids the use of coal coke in the production of pig iron by using sustainable charcoal instead. Greenhouse gas (GHG) emission reductions are achieved through afforestation and reforestation, charcoal production, and in the industrial process for pig iron production.

The project was started in 2001 and includes the establishment of 23 100 ha of high yielding Eucalyptus varieties to produce wood for charcoal, which will displace coke in the pig iron production. In addition, emissions reductions will be accomplished through the reduction of methane emissions in the charcoal production and regeneration of *cerrado*[1] native vegetation in 478 ha of pasture land. The total reductions expected along the project life are specified in Table 14.4.

The project is located in the Southeast of Brazil, State of Minas Gerais, in the municipalities of Sete Lagoas, Curvelo and Felixlândia. It evolves during twenty-eight years, in accordance with the 7-year Eucalyptus forest-growing cycle in the region. Beyond the climate change mitigation benefits, the project promotes the use of renewable resources and sustainable socio-economic development in the rural areas of Minas Gerais. The Plantar project follows the requirements for CDM projects and has been recently approved by the Prototype Carbon Fund at the World Bank.

[1] The *cerrado* is the native landscape in the area of the project. The *cerrado* covers large extensions in Brazil and is sometimes referred to as the Brazilian type of savannah. This native forest has been traditionally used for the production of charcoal.

Bioenergy – Realizing the Potential

14.2. OVERVIEW OF THE PIG IRON AND STEEL SECTORS IN BRAZIL

The Brazilian iron industry is divided into three main groups: (A) independent small pig iron producers; (B) large integrated steel mills using charcoal or coal coke; and (C) integrated coke-based steel mills. The total production of pig iron in 2000 was 27.7 million tons, of which 14 per cent were exported and the rest consumed in Brazil (IBS 2001). Table 14.1 shows the pig iron production in Brazil by producer.

The first group, small independent pig iron producers, in which the sponsor of the Plantar project operates, is the most numerous. There are about 40 independent firms producing pig iron in the State of Minas Gerais alone, with approximately 80 blast furnaces. These companies are mainly focused on the supply of iron to foundries and mini-mills in the international market and replacement of scrap used in electric arc furnaces of large steel producers in the domestic market (Group C as shown in Tables 14.1 and 14.2).

The plants of small independent producers usually have established capacity between 60 and 400 thousand tons per year, which is much smaller than those of the producers in the other two groups. Technological and economic constraints render unfeasible the conversion of small blast furnaces to use coke as has been the case among large producers. As a consequence, this sector remains dependent on charcoal as the main source of raw material for iron reduction.

The second group comprises four large private companies that originally produce steel mainly from charcoal: Acesita and Belgo-Mineira, which previously used both coke and charcoal as a reducing agent; and Mannesmann (now V&M do Brazil) and

Table 14.1. Pig iron production in Brazil (in thousands of tons)

Company group / Company	1995	1996	1997	1998	1999	2000
Group A: Independent Producers (total)	4919	4156	4564	4732	5169	5916
Group B: Large integrated mills	2460	2406	2362	2394	2356	2854
Acesita	471	496	562	587	623	685
Belgo-Mineira	770	777	697	769	696	935
Gerdau	701	621	622	625	669	722
V & M do Brasil	518	512	481	413	368	512
Group C: Large integrated coke mills	17642	17416	18087	17985	17024	18953
Usiminas	3929	3826	3738	3817	2851	4134
Açominas	2342	2286	2273	2260	2316	2538
Barra Mansa	82	–	–	–	0	0
Cosipa	3404	3427	3656	3369	2477	2748
CSN	4383	4358	4791	4561	4650	4517
CST	3502	3519	3629	3978	4730	5016
TOTAL in the three groups	25021	23978	25013	25111	24549	27723

Note: Group A refers to companies producing exclusively pig iron, and excludes cast iron tune producers.
Source: (IBS 2001).

Table 14.2. Steel production in Brazil (in thousands of tons)

Company group / Company	1995	1996	1997	1998	1999	2000
Group B: Large integrated mills	5586	6079	6293	6241	6677	7442
Acesita	612	624	632	687	786	856
Belgo-Mineira	1661	2054	2117	2157	2267	2571
Gerdau	2752	2878	3043	2964	3259	3496
V & M do Brasil	561	523	501	433	365	519
Group C: Large integrated coke mills	18 580	18 331	18 971	18 744	17 583	19 731
Usiminas	4160	4039	3930	4023	2980	4438
Açominas	2435	2400	2376	2330	2355	2620
Barra Mansa	308	351	364	346	390	393
Cosipa	3598	3604	3791	3519	2593	2746
CSN	4340	4364	4796	4708	4851	4782
CST	3739	3573	3714	3818	4414	4752
TOTAL	24 166	24 410	25 264	24 985	24 260	27 173

Source: IBS (2001).

Gerdau, both still reliant on charcoal. This type of enterprise, charcoal-based steel making, is only found in Brazil. Because of their production volumes (which range from 600 000 to 4.1 million tons of steel per year), they can make the conversion from charcoal to coke, which indeed has been accomplished by some companies. This group is, thus, the most responsive one to market factors in relation to the reducing agent (charcoal or coke) used.

The third group, integrated coke-based mills, currently account for 74 per cent of total pig-iron production in Brazil. This group comprises five large-scale enterprises (CST, Usiminas, Açominas, Cosipa, and CSN), previously parastatals that have been privatized since the early 1990s. The current trend has been towards an increasing concentration of the industry in integrated mills. These companies produce most of the pig iron produced in Brazil to supply their own steel mills (see Tables 14.1 and 14.2).

The forest legislation has forbidden the use of charcoal from native forests after 1992 (Forest Law of the State of Minas Gerais number 10.561 on 27.12.1991 and Decree 33.944 on 18.09.1992)[2]. Since then, the industries began to reduce the use of

[2] The laws concerned include also decree 1282, dated October 29, 1994, which states the rules for compliance with article 21 of the Brazilian forestry code (law number 4771, dated from September 15, 1965). The latter demands that the mills, freight companies and others, which are based on charcoal, fuel wood or other forestry raw materials, must keep their own plantations and explore them rationally. Supporting legislation there is normative instruction number 001, dated September 5, 1996, which provides rules on forestry reposition and law number 9605, dated February 12, 1998, which defines penalties to be applied for activities causing damage to the environment. These legal instruments are administered by IBAMA (Brazilian Institute for the Environment and Renewable Natural Resources).

Table 14.3. Consumption of charcoal in pig iron production (in thousand cubic meters and percentage of total)

Year	Charcoal from indigenous forests	%	Charcoal from planted forests	%	Total charcoal use
1988	28 563	78.0	8056	22.0	36 619
1989	31 900	71.2	12 903	28.8	44 803
1990	24 355	66.0	12 547	34.0	36 902
1991	17 876	57.0	13 102	42.3	30 978
1992	17 826	61.1	11 351	38.9	29 177
1993	17 923	56.5	13 777	43.5	31 700
1994	15 180	46.0	17 820	54.0	33 000
1995	14 920	48.0	16 164	52.0	31 084
1996	7800	30.0	18 200	70.0	26 000
1997	5800	25.0	17 800	75.0	23 600

Source: ABRACAVE (Brazilian Association of Renewable Forests).

charcoal from native forests and increase the consumption of charcoal from forest plantations. However, such charcoal was mostly produced from eucalyptus forests that were planted with tax incentives from the Federal Government between 1966 and 1986. Table 14.3 depicts the total decrease in charcoal utilization between 1988 and 1997, as well as the balance change between charcoal from indigenous forest vis-a-vis planted forests. Nonetheless, in spite of current legal and policy provisions, Brazil still faces great lack of fuelwood plantations and law enforcement in what concerns native wood usage. Even though local and state authorities have improved inspection and surveillance, charcoal manufacturing is often based on deforestation which also leads to intensive methane emissions.

According to statistics from IBAMA (Brazilian Institute for the Environment and Renewable Natural Resources), the stocks of existing forests can supply the market demand for charcoal for the next seven years. On the other hand, a recent study from BDMG (Development Bank of State of Minas Gerais) indicated demand for 110 000 ha of forests in 2007 while no forests are being planted to supply charcoal markets. In fact, regulatory, operational and economic barriers are leading to the decline of wood-plantations which contrasts with the effective demand for charcoal, and increases pressure on native wood sources.

Eucalyptus forest plantations in Brazil need approximately seven years to mature. Thus there is the need to plan in advance to guarantee the survival of the charcoal producers on a sustainable basis. Forest plantations with selected clones and high productivity, aimed at re-establishing the stocks of charcoal can have a positive effect in changing current trends and foster a larger use of charcoal in pig iron industries. This can be done following the environmental sustainability principles while also guaranteeing employment for many charcoal producers.

14.3. BASELINES

The Plantar project is based on three distinct but inter-linked baselines, and an additional pilot project of rehabilitation of native vegetation as follows:

- The forestry component, which involves the establishment of new Eucalyptus plantations;
- The carbonization process component, which involves altering the design of carbonization kilns and/or the carbonization process to reduce the emissions of methane;
- The industrial component, which consists of the avoidance of coke in the production of pig iron by using charcoal from sustainable forests;
- The pilot project on rehabilitation of native vegetation in the Brazilian *cerrado*.

The baseline scenario for the project is based on what would have happened in the absence of the project activities. In this case, the baseline scenario would be the gradual death of the charcoal-based independent pig iron producers, giving place to increasing production based on coal in large mills. The latter would gradually take over the market share controlled by small pig iron producers. In addition, the existing planted forests will not be replanted after the third cycle of rotation and, most probably, the land will be used for pasture, which is the most common activity in the region.

The Plantar project scenario allows for the use of sustainably produced charcoal as the reduction agent in the production of pig iron, avoids the absorption of the charcoal-based market share by the coke-based pig iron production, and also avoids GHG emissions. The CDM framework allows the project activity to be cofinanced through the sales of carbon credits, protecting from the potential loss of the industry and market shares of small charcoal-based pig iron producers. Moreover, the forestry activities to supply charcoal to Plantar's mills are based on new high yielding plantings.

Figure 14.1 shows a comparison of two routes to produce pig iron and demonstrate the main steps of the production process utilising coke and charcoal. In both cases, there are greenhouse gas emissions in the carbonization process. The types of kilns used by Plantar (currently approximately 2000 kilns) and other charcoal manufacturers in Brazil produce emissions of methane. Plantar will invest in the refurbishment of its kilns, and in the improvement of the carbonization process to reduce the emissions of methane to the atmosphere.

Finally, Plantar will initiate a pilot project to manage some of its lands to enable the regeneration of the *cerrado* and other native vegetation in lands that were previously planted with Eucalyptus or were used as pasture. If successful,

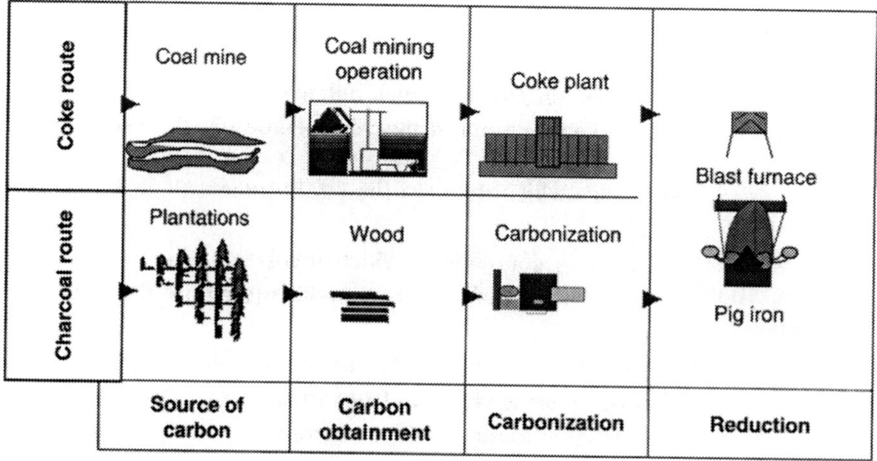

Figure 14.1. Comparison of coke and charcoal-based pig iron manufacture.

Table 14.4. Plantar project cumulative baseline emissions and emissions reductions (tons of CO_2)

Year	Total emission reductions (estimated in 28 years)			Total emission reductions
	Forest sequestration	Carbonization methane emission reductions	Industrial processes (pig iron production)	
2001–2028	4 545 398	437 325	7 903 262	12 885 985

Plantar intends to expand this component of the project to other lands elsewhere. The balance of emissions reductions expected from the whole project is shown in Table 14.4.

14.4. PROJECT BOUNDARIES AND LEAKAGE

The project's boundaries were assumed to be the Brazilian territorial boundaries. Most sources of emissions and emissions reductions associated with this project, which take place in Brazil, were accounted for. This includes transport of charcoal and coal/coke that will or would take place in the project and baseline scenarios, as well as the coking process that is expected to take place in Brazil. The emissions associated with coal mining elsewhere and coal transport to Brazil were not included in the analysis. There are good reasons for that.

First, if these emissions were included, it would be also plausible to consider the possible sources of transnational leakage potentially generated by the project. While one could argue that the project has led to a reduction in global consumption of coal (and this is picked up by the emissions reductions taking place in Brazil), one could also argue that the reduction in imports of coal to Brazil could lead to a reduction in international prices leading to increased consumption elsewhere.

Second, there is still a lack of definition regarding the "ownership" of emissions (and consequently emissions reductions) associated with international transport. The complexity of this type of analysis, associated with the lack of definitions regarding international "property rights" related to emissions reductions, were determinant in limiting the boundaries of this analysis to the territory boundaries of Brazil.

The mitigation activity of the Plantar project is unlikely to result in significant amount of leakage. While the coke-avoidance component of the project is based on the reduction of fossil fuel consumption (imported coal), thus in a sense making this amount of coal available to the rest of the world, in global terms the project is relatively small and unlikely to have any effect on the price and consumption in the global coal market.

The forestry component is based on the establishment of 23 100 ha of sustainable and high yielding forest plantations *in lieu* of pastureland. Consequently, no leakage is expected. On the contrary, it is expected that the project will result in positive offsite impacts, reducing the pressure on native forests for commercial charcoal production and for fuel wood.

In the past, charcoal producers from the State of Minas Gerais have looked for other areas to relocate their production, responding to the constraints imposed on the use of native forests for charcoal. In particular, the Carajás region in the Eastern Amazon has been a major target. In theory, this could be a possible source of leakage for the Plantar project. However, investment and environmental constraints in Carajás today are similar to those in Minas Gerais. Thus, this possibility of leakage seems quite unlikely.

From a commercial perspective, these small industrialists lack capital and collateral to borrow as their equity is trapped in Minas Gerais blast furnaces, which cannot be sold in the present market. Brazilian credit markets have been quite restricted, especially after the events of 11 September in the US.

The project requires careful monitoring of other small independent pig iron producers and the large integrated mills in Brazil as a whole. Data gathered from a reputable and independent source on trends in the iron and steel industry will be used to determine whether and to what extent the independent sector of pig iron producers is expanding production based on charcoal in comparison with the coke-based production, without the benefit of carbon finance. These data will contribute to the initial verification of the industrial coke-avoidance component of the project when it is commenced in 2008. It is presumed that maintaining and assessing the significance of these data will maintain the integrity of the baseline scenario.

Under the sequestration component, the project sponsor intends to claim carbon credits from established plantations on land that has been pasture land since 1989, taking into account the CDM Modalities and Procedures.

14.5. ENVIRONMENTAL ISSUES

The Plantar project will lead to total emission reductions of almost 13 million tons of CO_2. Table 14.4 indicates how much is the accomplishment in each component of the project.

One of the main criticisms of plantation forestry is related to biodiversity suppression. In the Plantar project, a number of precautions are taken to bring benefits in relation to biodiversity (Kornexl, 2001).

- The plantations are certified to the standards of the Forest Stewardship Council, a strict environmental standard related to sustainable forestry worldwide. The standard requires forest operations to ensure the maintenance of biodiversity within managed areas.
- Plantation-based and sustainable charcoal production reduces pressure on native forests. Currently, Plantar itself still uses charcoal from native forests in its pig iron mill (derived from legal and authorized deforestation conduced by third parties outside Plantar's own land). With the development of the project, Plantar will become fully self-sufficient in charcoal.
- The pilot project area of regeneration of *cerrado* and other native vegetation will lead to an increase in biodiversity and the return of native species of plants and animals. Plantar is also considering a more active management with assisted regeneration of the biomass, exploring landscape level biodiversity management opportunities.
- According to the baseline biodiversity study, fire suppression is the single most important biodiversity benefit of the Plantar project. By continuing its current fire monitoring and control system, Plantar could allow the *cerrado* and other native vegetation ecosystems on its land holdings to partially recover their original species composition through the process of secondary succession. Additionally, Plantar already provides neighboring landholders with the benefit of fire watch towers, which expands the positive impact of the project (Nepstad and Vale, 2001).

Six indicators to follow up on the biodiversity benefits accrued from the Plantar project are suggested in the study by Nepstad and Vale (2001):

- The total area of the legal reserve in the Curvelo property beyond 20 per cent (the likely area as of project initiation);
- The total area of fragments larger than 50 ha beyond the 2002 baseline;
- Reduction in fire incidence within 10 km of the ranch relative to regional fire incidence (10 to 30 km distant), using a 1999–2001 baseline of fire incidence;
- Number of native species of birds per sampling effort relative to 2002 baseline in three legal reserve fragments;
- Biomass increase in native vegetation beyond 2002 baseline;
- Testing of Eucalyptus effects on stream-flow, and incorporation of watershed management principles into harvesting regime, if warranted.

In addition to the efforts on biodiversity, Plantar will continue monitoring water hydro-biological, physical and chemical quality, and building up corridors between remaining native forest fragments, and recuperating former deforested areas in environmental fragile zones. The monitoring activities aim at accompanying the development of water quality conditions and seasonal variations, and the impact of forest activities on water quality. This information will help in identifying eventual shortcomings and propose measures to reduce impacts as well as contribute to evaluate the efficiency of the implemented environmental management system.

14.6. SOCIOECONOMIC ISSUES

The pig iron and carbonization segments are important sources of employment in Brazil. By 1998, the pig iron segment employed 131 000 people in Brazil, about two-thirds of these in Minas Gerais. The vast majority was employed in reforestation and production of charcoal from native or planted sources (May and Chomitz, 2001). About 25 per cent were engaged in forward linked activities in the steel and foundry industries that use pig iron as an input (ABRACAVE, 2001). Charcoal production is the most labor-intensive part of the charcoal-based iron industry. Of the 84 000 employed in the independent pig-iron segment, 70 000 or 83 per cent are engaged in field activities related to cutting, transporting and carbonising fuel wood (SINDIFER, 2000).

Traditionally, labor conditions in the production of charcoal have been appalling. The field activities related to cutting, transporting and carbonizing fuel wood have been historically criticized for health and safety risks, poorly remunerated labor conditions and child labor. Growing social awareness and concern has led to a worldwide campaign against rural wage slavery and child labor in Brazil (May and Chomitz, 2001).

The Plantar Group (which also includes a division for provision of services to third parties) currently employs 5500 people, who may retain their jobs as the project goes ahead. Labor conditions in Plantar are above average and the firm does not employ children, as certified independently by the SCS Group as part of the Forest Stewardship Council's certification process. The planned changes in the carbonization process are also expected to lead to substantial improvements in the health of employees. Associated with the social requirements of the Forest Stewardship Council, there is scope for Plantar to provide a better model of socially responsible enterprises.

The pig iron sector is beneficial to the Brazilian economy, being a major employer, responsible for significant amount of exports (US$ 445 million FOB in 2000 according to IBS 2000), and the source of raw materials to other industries in the country.

The Plantar project provides a new model for financing the charcoal-based pig iron industry in Minas Gerais and Brazil, allowing for the survival of independent producers and the plantation forestry sectors in the region. This new business model could also help attract substantial additional foreign investment to the country, with positive effects to the Brazilian balance of payments.

Considering the focus on the small independent producers, there are also important benefits to be accrued from wealth distribution and development of small and medium sized enterprises. The multiplier effect of this investment is likely to bring additional benefits, particularly in rural areas where the project is located. It will result in additional job creation and preservation of jobs associated with forestry activities, having important effects in the regional rural economy.

REFERENCES

ABRACAVE (2001) Anuário Estatístico, Yearbook of the renewable charcoal industry, www.abracave.com.br.

Eco Securities, Prototype Carbon Fund (2002) Baseline determination for Plantar: Evaluation of the emissions reduction potential of the Plantar Project.

IBS (2001) Anuário Estatístico 2001, Brazil Steel Databook.

Kornexl, W. (2001) Environmental Assessment of Plantar Project, Minas Gerais, Brazil.

May, P. and Chomitz, K. (2001) The charcoal-based iron industry in Minas Gerais, Brazil, and the global environment. World Bank Development Research Department.

Nepstad, D. C. and Vale L. C. C. (2001) Biodiversity Benefits of the Plantar Carbon Project: A proposal for Baselines and a Protocol for Monitoring and Verification (MVP), World Bank Prototype Carbon Fund, PCF internal document.

Prototype Carbon Fund (2002) Brazil: Sustainable Fuelwood and Charcoal Production for the Pig Iron Industry in Minas Gerais, The Plantar Project, Project Design Document.

SINDIFER (2000) Annual Statistics report, Yearbook of steel industry statistics, http://www.mme.gov.br/smm/anuario2000/anuario.thm.

Chapter 15
Carbon Credits from Cogeneration with Bagasse

Marcelo Junqueira

15.1. THE CONTEXT OF SANTA ELISA'S BAGASSE COGENERATION PROJECT

The demand for electricity in Brazil is growing fast, requiring new and cost-efficient ways to meet the country's energy needs. Brazil's predominant reliance on hydropower (80 per cent of the country's total electricity use in 2002) is expected to decrease as diversification is being pursued. The competitiveness of large hydropower falls in face of relatively high construction and transmission costs. In addition, about half of Brazil's remaining hydro potential is located in the Amazon area, where many social and environmental constraints hinder further development.

In response to that, Brazilian authorities have developed a thermoelectric program. The Brazilian Ten-year Expansion Plan for 1999–2008[1] counts on increased involvement of private capital in the electricity sector for the construction of new thermal plants. The thermoelectric expansion plan is based on the use of natural gas, mineral coal and, in the case of isolated electricity systems, petroleum derivatives. In the South and Southeast regions (where Companhia Energética Santa Elisa is located) only natural gas will be used.

Within the above scenario, the amount of greenhouse gas emissions are expected to increase in Brazil. Meanwhile, the Kyoto Protocol and, in particular, the Clean Development Mechanism (CDM) provides a useful financial tool to foster the implementation of new solutions to supply the country's energy demand, while also avoiding the increase of greenhouse gas emissions.

This chapter presents an actual case of mitigation of greenhouse gas emissions through the provision of renewable energy utilising biomass residues (bagasse) from the sugar and ethanol production at Companhia Energética Santa Elisa. Santa Elisa is the third largest sugar producer in Brazil and also the third largest electricity-producing sugar mill through cogeneration[2]. The project results from Santa Elisa's

---

[1] Electrobrás and the Brazilian Ministry of Mines and Energy, 1999. Ten-Year Expansion Plan: 1999–2008, produced by the GCPS Electric Systems Planning Coordination Group, Brazil.

[2] Of approximately 300 sugar mills in Brazil, less than half sell surplus electricity to the grid. The majority of mills produce energy solely for on-site use, which is the Business-as-Usual for the sugarcane industry.

decision to expand its cogeneration system to increase efficiency and aggregate value to the bagasse originated from its sugar milling process. Santa Elisa intends to validate its investment within the CDM. An agreement has already been signed with the Swedish government concerning the certificates that will accrue from the project.

The Brazilian Inter-Ministerial Commission on Global Climate Change, which is responsible for defining national eligibility criteria for the application of CDM in the country, has determined that renewable energy cogeneration projects meet the sustainable development criteria sought by the Brazilian government. In this context, both sugarcane-based products and electricity provision are considered essential for the sustainable development of Brazil. Thus, the Santa Elisa project has received official support.

15.2. COGENERATING WITH BAGASSE – THE PROJECT MILESTONES

In 1993, Santa Elisa's cogenerating capacity was less than 13 MW. Until then, the company had focused on meeting its own energy demand only. In 1993, the power plant was enlarged and two 8 MW turbo generators were added. In 1998, the company shifted an old 4 MW turbo generator to a new 6 MW. As a result, the total installed capacity in 1999 was 31 MW. From 1994 until 2001, Santa Elisa was able to supply the grid with approximately 5 MWh during the harvest season which, in this region, serves to meet the needs of a town with approximately 60 000 inhabitants.

Santa Elisa's bagasse cogeneration project involves new investments to install new boilers and generators, and increase capacity and efficiency of the plant. At the same time, investments in the production of sugar and alcohol are being made, so that the steam consumption of the sugar production process is reduced to 400 kg of steam per ton of sugarcane crushed. The phases of the project are described below.

Phase 1 - 2003

A high efficiency pressure boiler providing 65 bar at 200 ton of steam per hour and 510°C (the first in the sugar industry in Brazil) is installed. This implies a significant reduction of the amount of bagasse used per ton of steam generated. Two new contra pressure-type turbo generators at 15 MW each and two new condensing-type turbo generators at 6 MW are also installed. Other investments projected include the construction of a new powerhouse, a new sub-station with a new measurement equipment, and a transmission line. Also the bagasse deposit is doubled, reaching a capacity of 80 000 tons of bagasse, equivalent to a production of 8 MWh during 6 months.

Table 15.1. Installed capacity in Santa Elisa bagasse co-generation project

	1999 (base for credits)	Phase 1 2003	Phase 2 2004	2005	2006	2007	2008	2009
Installed Capacity, MW	31	58	73	73	73	73	73	73
Internal Consumption MW	20–23	20–23	20–23	20–23	20–23	20–23	20–23	20–23
Capacity to be exported (Forecast)** MW	5*	35–38	50–53	50–53	50–53	50–53	50–53	50–53
Capacity applying for credits MW	Not Applicable	25	45***	45***	45***	45***	45***	45***

* Average amount of energy being exported in the years before the investment for energy producing and selling in 2003, as a marginal cogeneration of energy according to fuel availability and sugar mill consumption

** The Forecast is made upon PPA considerations and market expectations. For 2003 the PPA requires a supply of 30MW and provides CPFL with a buying option of 3MW, the rest can be sold on the spot market if SE wants so.

*** Forecast of firm energy sales minus average energy sold before the investment in 2003, considering that the energy sold before was a marginal generation dependent on the availability of bagasse.

Santa Elisa deactivates two old generators and three less efficient 21 bar boilers. The production capacity increases to 58 MW. With these investments, Santa Elisa achieves an installed capacity between 30 MW and 40 MW to produce electricity to the grid during the harvest season of 2003 (see Table 15.1). Investments reach US$ 20.7 million in this first phase. For the local utility company, it is advantageous to buy energy produced by a sugar mill, as the base load for utilities in Brazil is supported mainly through hydro generation, and the sugarcane crop season coincides with the dry period. CPFL[3] has signed a ten-year purchase contract with Companhia Energética Santa Elisa.

Phase 2 - 2004

During the harvest season of the year 2004, higher capacity and efficiency is reached through the acquisition of another 15 MW turbo generator, another 65 bar boiler, again targeting a steam consumption at 400 kg per ton of sugarcane crushed. The forecast is to produce around 216 GWh of clean energy annually to supply the regional grid.

Second phase investments are expected to reach approximately US$ 13 million. The investment to increase efficiency is dependent also on the expansion of sugar production. The financial support from certified emissions reductions (CERs) is

[3] CPFL is a leading energy distributor in the State of São Paulo.

helpful in improving the mill competitiveness and enhancing the sustainability of the project as a whole.

Considering that the cogeneration depends on the biomass supply to the sugar mill boilers if, for some reason, agricultural operation or transportation is disrupted, the sugarcane will not reach the sugar mill and the boilers will not be able to produce the steam required. For this reason, the expansion plans for energy generation at Santa Elisa come together with significant investments in the sugar production process. The idea is to reduce steam consumption in the sugar production process to release as much as possible to cogeneration.

Sources of funding

BNDES, the National Bank for Economic and Social Development, is financing approximately 80 per cent of the R$ 48 million (US$ 20.7 million) investment for the first phase of the project, following the technical specifications described above. This is done through the Special Agency for Industrial Finance (FINAME) as shown:

Beneficiary:	Companhia Energética Santa Elisa
Value:	R$ 35 million (US$ 15 million, April 2002)
Period:	10 years, including 2 years of grace period
Interest rate:	3.5 per cent + TJLP[4], including 1.5 per cent risk spread

The funding sources for the second phase are of the order of US$ 13 million to increase the installed capacity to 20 MW but had not yet been secured at the time of this writing.

The total volume of carbon credits expected to be generated over the seven-year (2003–2009) crediting period is 635 501 million tons of CO_2. Even though the economic impact of the certificates on the investment is relatively small from the investors' point of view, it is a means of improving the rank of the proposed project against other competing investment options. It helps to make renewable energy more competitive against fossil fuel plants such as natural gas combined cycle plants, which are very competitive and, currently, the business-as-usual case in Brazil. Thus the CDM framework brings the cogeneration expansion plan closer to any other alternative investments of equal risk.

15.3. ADDITIONALITY

The Climate Convention requires carbon offsets based on certified emission reductions (CERs) to be clearly additional. This criterion demands that selected projects

[4] TJLP is the Brazilian long term interest rate.

have a credible, quantifiable and verifiable baseline of emissions, from which reductions can be measured and verified. The baseline represents the emissions from electricity generation that would occur in the absence of the certified project activity. A clearly additional project is one that represents actions that would have little chance of being taken without the use of the CDM. The reason why an independent body is needed to certify emissions reductions is that an offset transaction is not totally straight forward. Both the buyer and the seller could benefit from exaggerating the emissions reductions. To ensure that total emissions indeed decrease, emissions reductions must be real and measurable in reference to a defined baseline.

The goal is to acquire carbon offsets (CERs) of high quality. The quality of CERs from an offset project depends on the credibility of the project's additionality. Thus, "the baseline describes the greenhouse gas emissions associated with a counterfactual scenario that would prevail without the JI or CDM intervention and with which actual emissions can be compared" (World Bank, 1999a). The credibility of the baseline is crucial, as this is the key to the acceptance of the project's CERs.

Reliable supply of electric power is a key input for the industrialization process of developing countries' economies. In Brazil, the growth rate of this sector is higher than that of the overall economy, as electrification is often closely linked to development priorities. The growth rate of energy demand in Brazil drives the government to invest in ready-to-use technology instead of developing new alternatives even if they could result in less greenhouse gases per MWh generated. The government's expansion plan for the energy sector pushes ahead the thermal energy generation from 9 to 17 per cent of the installed capacity from 2001 to 2004. In this context, the carbon intensity of the electrical system obviously increases.

A cogeneration project based on renewable resources such as Santa Elisa is environmentally additional as it contrasts with what is the business-as-usual and thus likely to happen in the absence of the project. Therefore, the project is eligible under the CDM and can generate CERs. The updating of the multi-project baseline at regular intervals will be important to ensure that developments in the electricity sector are captured in the assessments of the project.

15.4. PROJECT BASELINES

The question of whether emissions reductions are additional to what would have been achieved without the offset project depends on the counterfactual conditions without the project, i.e. the baseline. The baseline emissions are the greenhouse gas emissions expected in the absence of the proposed project. Because the credibility of an offset depends on additionality, it requires a quantifiable and verifiable baseline of emissions. Thus the establishment of the baseline is key in determining the extent to which a carbon offset project is additional under the CDM.

The expected timing of emission reductions or carbon storage benefits can depend on the dynamics of the baseline or reference case. To the extent that the baseline case involves a pattern of emissions over time, the earlier those emissions occur and can be reduced by a carbon-offset project, the sooner the project can claim offset credit. This timing can strongly influence the economic performance and risk exposure of a project. For this reason, the baseline needs to be updated during the project life, and the procedure for updating the baseline forms part of the monitoring and verification protocol for the project.

In order to reduce the controversy regarding additionality and baselines, specific criteria for establishing project baselines are needed. Ideally, baseline criteria should be universal, but the potential range of CDM projects is too diverse. The criteria for baselines may vary geographically across different countries and regions, as well as technologically across different sectors and types of projects.

Thus the baseline is not static, and time variations in the generation fuel mix must be captured in the baseline carbon intensity. This is where a benchmark approach can be used to simplify the analysis. The simplest benchmarks for baseline emissions from electricity generation are (i) to use the average emission rate for the entire system (i.e. total emissions divided by total sales) or (ii) to use the weighted-average marginal emission rate.

For the Santa Elisa project, the appropriate benchmark for the baseline is the second method, that is *the weighted-average marginal emission rate*. The reason for this is simply the logic that governs dispatching in a generation system dominated by hydroelectric sources. Hydro resources will always be dispatched as much as possible first. Thermal sources are dispatched only when necessary to meet larger loads.

Thermal sources are a significant component of the baseline case for the Santa Elisa project. Although the total generation mix will still be dominated by base-load hydro sources, most of these sources would operate either with or without the Santa Elisa project. Considering the emissions (zero) of these base-load hydro sources in the baseline carbon intensity, it would be misleading to use the average emission rate benchmark.

In order to proceed with the quantification of the baseline scenario for the project, we need to specify the basic principles to be followed in the baseline scenario. These principles should be guided by the discussion of additionality in the CDM, and it should be adapted to the actual situation in the electric power sector of Brazil. Once we have stated the principles for defining the baseline scenarios, we can explore the detailed analysis of electricity generation dispatching and expansion planning in order to identify the baseline generation sources. Then we can determine the corresponding baseline carbon intensities against which the project should be compared to determine net emission reductions. Finally, we need to consider the updating of these estimates in the future, as an input to the monitoring and verification plan for the project.

For the Santa Elisa project, we have used a baseline methodology in which both the current Brazilian energy system and the government expansion plans are included. In this context, we have taken into consideration the economic attractiveness of thermal plants in Brazil and the shift to private-sector financing which favors thermal sources and is less conducive to hydropower. As a result, the profile of new generating capacity is likely to be different from the existing installed capacity, and power-sector expansion shall include natural gas combined-cycle thermal stations. Renewable generation sources implemented with private-sector investment should, therefore, be considered "additional to any that would occur in the absence of the certified project activity."

Using the method described, we can estimate the carbon emission intensity of the baseline generation that will be replaced by the output of the Santa Elisa project. During the lifetime of the project, this baseline intensity will depend on the type of thermal generating stations installed at the margin in the Brazilian system and their emissions rate. The specific details of the baseline sources might change and make adjustments necessary in relation to the baseline carbon emissions intensity values. We have selected a crediting period starting in 2002 for a maximum of seven years, which may be renewed two times.

15.5. QUANTIFYING BASELINE CARBON INTENSITY

Santa Elisa will sell energy to the Companhia Paulista de Força e Luz (CPFL). CPFL is an electric distribution company serving 234 municipalities in the State of São Paulo. The company has had an annual growth of 4.8 per cent on average between 1989 and 1998. The peak demand of CPFL was approximately 5000 MW in 2001. CPFL's self generation amounted to only 2.7 per cent of its total energy sales in 1995, and came mainly from hydro plants: 112 MW of hydro and 36 MW of diesel oil-fueled installed capacity[5] (e.g. Carioba). CPFL's main supplier is CESP (Companhia Elétrica de São Paulo), from which CPFL purchases 95 per cent of the power, essentially all hydroelectric. CPFL must also buy 600 MW from Itaipu hydro station. In addition, CPFL has been planning a new gas-fired combined-cycle plant of 350–400 MW near Campinas next to the natural gas pipeline.

The existing installed capacity of CPFL (self-generation plus purchases under bilateral contracts) was able to meet the demand in 2001 due to the energy rationing

[5] Cogeração Setor Alcooleiro, CPFL, 1995.

imposed by the federal government. Specialists estimate that the energy rationing in 2001 shrank the energy market by approximately 6 per cent. In conditions of demand growth of 3.5 per cent, however, CPFL would have had to purchase energy in the spot market to meet its demand already in 2001 as the company did not have the required energy negotiated under bilateral contracts to meet the whole demand.

Since 85 per cent of the distributors' purchases in the free market must come from bilateral contracts, and assuming that some sales are diverted away from the distribution utilities by direct sales to customers, we can assume that a somewhat smaller share (80 per cent approximately) of total generation is sold in the free market through bilateral contracts. The remaining 15–20 per cent would be negotiated in the spot market. The spot market sales are estimated to be the company's total generation in an average hydrological year, minus the sales to initial contracts and bilateral contracts.

In the spot market, all the so-called South/Southeast system loads are intended to be met with the least-cost combination of available resources, comparing the value of water in an average hydrological year with the variable cost (fuel plus O&M) of the marginal thermal plant. The marginal plant is gas-fired. Between 2001 and 2003, 5.4 GW of new gas-fired capacity were added only to the S/SE and Midwest regions of Brazil[6]. Thus, the marginal plant for the Santa Elisa project baseline is gas-fired.

As a consequence of the rationing and nonliquidation of the spot market in 2001, the CPFL did increase the use of its old (1953) diesel oil-fueled thermal plant, Carioba, with 36 MW of installed capacity. One could argue that the fuel replaced in the first year of operation of the Santa Elisa project should be diesel oil, as used in Carioba, with low net efficiency. However, the project participants decided to be conservative and estimate the CERs originated by the project based on the marginal energy sources available at the spot market, where CPFL would have purchased energy otherwise. A calculation of emission intensity based on gas at the margin is an adequate baseline benchmark for small generation projects that do not cause changes to the generation expansion plan.

Changes to the generation expansion plan would make the marginal source more difficult to identify for the purpose of calculating emission changes. This is because the resulting emission change might be savings from a hydroelectric plant or a fossil-fired plant that would be deferred or completely displaced from the future generation mix, and that would not resemble the average or the marginal resource at all. Thus, defining a credible baseline case entails analyzing the existing expansion

[6] Programa Estruturado de Aumento da Oferta 2001–2003, Governo do Estado de São Paulo, Secretaria do Estado de Energia.

plan, for the entire national grid, to determine the generating resources that would be replaced by the CDM project, in this case Santa Elisa, and the emissions from these electricity-supply resources.

Thus, to establish the benchmark, we examine the new capacity additions called for in the recent versions of the national expansion plan. The objective is to characterize the potential new sources that could be deferred or replaced by the Santa Elisa project. Based on this approach, we can estimate the emissions of the incremental capacity displaced by Santa Elisa using its carbon emission intensity, weighted according to generation from the project.

15.6. CARBON ACCOUNTING EVALUATION METHODS

This section suggests specific steps to quantify net emissions reductions at Santa Elisa. The objective is to arrive at a conservative estimate established in metric tons of carbon-equivalent (mtC). As previously mentioned, net emissions reductions for the Santa Elisa project must be compared on the basis of the carbon content of the fossil fuel replaced. Thus, once a credible baseline has been identified, the principal parameters that determine the actual emissions reductions are (i) the baseline carbon emission intensity, (ii) the project emissions (if any) and (iii) the projects' energy production rates.

Once the baseline case has been defined, the carbon accounting for energy supply projects is relatively simple. Since the Santa Elisa project will result in electric energy generation from biomass renewable sources, the net greenhouse gas savings will be realized from the reduction of fossil fuel use in thermal power generation plants that would supply electricity in the baseline case. The Santa Elisa output is then multiplied by the appropriate carbon intensity for the associated baseline electricity to determine net greenhouse gas emission reductions as follows:

$$ER_{\mathrm{net}} = E_{\mathrm{b}}C_{\mathrm{b}} - E_{\mathrm{p}}C_{\mathrm{p}}$$

where:

E_{b} = Energy produced in baseline case (generally assumed to be equal to E_{p})
C_{b} = Carbon intensity of energy in baseline case
E_{p} = Energy produced in the project case
C_{p} = Carbon intensity of energy in project case (generally assumed to be zero)

In many renewable energy projects, one can generally assume the project carbon intensity is zero, including sustainably grown biomass fuel. Thus, the carbon emission reduction is the product of the baseline carbon intensity and the measured energy supplied (or sold) by the project. Thus:

$$ER_{\text{net}} = E_p C_b - E_p C_p = E_p C_b$$

These two remaining values, the baseline carbon intensity (C_b) and the electric energy produced by the projects (E_p), are the parameters that must be quantified and measured in order to generate certified emission reductions (CERs).Thus, renewable energy supply projects can be relatively simple in that they require monitoring of only the project emissions (if any) and the energy production rates, once the baseline carbon intensity has been determined.

The baseline carbon intensity (C_b) is based on the carbon content of the fuel combusted by the baseline generation source and the efficiency with which that source operates. The assumed baseline source burns natural gas. The standard factor for the carbon content of natural gas is 0.0153 mtC/GJ according to the IPCC (1996). The typical marginal source is a combined-cycle turbine plant with a net thermal efficiency of 45 per cent, which corresponds to a heat rate of 8.0 GJ/MWh. From these two parameters, the baseline carbon intensity can be calculated as follows:

$$\text{Carbon intensity (mtC/MWh)}$$

$$= \frac{3.6\text{ GJ/MWh [by definition]} * \text{Carbon content of fuel (mtC/GJ)}}{\text{Net thermal efficiency of plant}}$$

$$\text{Carbon intensity (mtC/MWh)} = \frac{3.6\text{ GJ/MWh} * 0.0153\text{ mtC/GJ}}{0.45}$$

$$= 0.122\text{ mtC/MWh}$$

The renewable generation technologies emit little or no direct GHG emissions. To the extent that there are some direct CO_2 emissions from fossil fuel use, for example for start-up or stand-by generators or for biomass fuel production and transportation, these emissions should be deducted from the total project emission reductions. However, there is no need for separate baseline calculations to account for this.

The primary source of net emission reductions for renewable energy projects is the reduction in fossil-fuel use at thermal generating stations that can be replaced or deferred by the project. This emission reduction mechanism is generally the only relevant mechanism. However, one should consider other emission sources and reduction mechanisms that have been identified in the biomass energy project.

These mechanisms include non-CO_2 GHG emissions, carbon sequestration, and indirect emissions resulting from the project.

At Santa Elisa, these emissions will be small or negligible, and we do not expect them to affect the baseline. Moreover, the fuel sources for the biomass energy project are assumed to be in the form of residues rather than wood from forestry plantations. Nonetheless, use of biomass fuel produces no net emissions, if the biomass is produced sustainably within the project.

Indirect emission impacts can result from project construction, transportation of materials and fuel (at least in the baseline), and other up-stream activities. These emissions are expected to be negligible compared to the emission reductions resulting from replacing thermal generation with renewable sources. These up-stream activities are also outside the system boundary, which includes the Santa Elisa Sugar Mill, the existing electricity generation and transmission system, and the future generation and transmission facilities to which the project will be interconnected.

15.7. LIFETIME OF THE PROJECT

The total number of years for which a multiproject baseline will be considered adequate to reflect "what would occur otherwise" is the key to determining the amount of emission units that could be expected from the CDM project. Determining up-front the crediting associated with a multiproject baseline would also enhance the transparency and consistency of the project, in addition to providing some certainty for the project's potential CER buyers.

A recent Dutch study on baselines suggests considering the development of a generic list of time horizons based on the type of project (JIRC, 2000). The Brazilian climate change experts have suggested using lifetimes of 15 to 50 years for different types of power plants in Brazil, with 15 years for gas turbines and internal combustion engine plants, and 25 years for steam turbines. For this specific project, the lifetime considered is the same as the Santa Elisa's investment program, which is set at 25 years. Some argue, however, that the carbon project lifetime should be shorter than the typical lifetime of power project investments.

Considering the aforementioned issues, the Santa Elisa Carbon Credit Project is aggregating the carbon credits generated in the expansion cogeneration project, which will deliver energy to the CPFL grid under a 10-year contract. In a conservative way, the project participants suggested a crediting period of 7 years. According to decisions from COP-7, it will be possible to renew the crediting period up to two times, provided that, for each renewal, a designated operational entity determines and informs the executive board that the original project baseline is still valid or has been updated. This diminishes the possibility of selling a nonexistent emissions reduction due to baseline changes.

Following on requirements of the Brazilian law, Santa Elisa has already received the Previous Environmental License from the Environment State Secretary (Secretaria de Estado do Meio Ambiente) and the Installation License from CETESB (Companhia de Tecnologia de Saneamento Ambiental), and can move ahead with the project.

REFERENCES

Econergy International Corporation (2002) Baseline Analysis and Quantification of Net Emissions Reductions from the Vale do Rosário Biomass Cogeneration Project.

E4 (1998) Using Area-Specific Cost Analysis to Identify Low Incremental-cost Renewable Energy Options: A Case Study of Co-generation Using Bagasse (Sugar Cane) in the State of São Paulo.

E4, Econergy International Corporation (2001) Procedures for Baseline Analysis and Quantification of Net Emissions Reductions from Renewable Energy Projects in Costa Rica.

International Energy Agency (IEA) (1997) Activities Implemented Jointly: Partnerships for Climate and Development, OECD, Paris.

Intergovernmental Panel on Climate Change (1996) Greenhouse Gas Inventory, Reference Manual, **Vol. 3,** Chapter 1, pp 1–13.

IPCC (1996) Guidelines for Natural Greenhouse Gas Inventories: Reference Manual, Energy.

JIRC - Joint Implementation Registration Centre (2000) Setting a Standard for JI and CDM: Recommendation on baseline and certification based on AIJ experience, commissioned by the Dutch Ministries of Economic Affairs, Foreign Affairs and Housing, Spatial Planning and Environment, The Hague.

Michaelowa, A. (1999) Template for baseline studies for World Bank AIJ, CDM and JI projects, Study for the World Bank Climate Change Team, December.

Swisher, J.N. (1997) Joint Implementation under the U.N. Framework Convention on Climate Change: Technical and Institutional Challenges, Mitigation and Adaptation Strategies for Global Change, **Vol. 2,** pp 57–80.

Swisher, J.N. (1998) Project baselines and additionality in the clean development mechanism, *Proceedings of the Aspen Global Forum – Post-Kyoto Strategies for International Cooperation and Private Sector Participation*, Institute for Policy Implementation, University of Colorado, Denver, October.

UNFCCC (2001) The Marrakesh Accords & the Marrakesh Declaration.

U.S. Dept. of Energy (DoE) (1995) *Proceedings of the United States Initiative on Joint Implementation (USIJI) Program Conference*, Arlington, VA, June.

World Bank (1999a) Baseline Methodologies for PCF Projects, The World Bank, Washington, October.

World Bank (1999b) Validation, Verification and Certification for PCF Projects, The World Bank, Washington, October.

Chapter 16
Wood Waste Cogeneration in Kumasi, Ghana

Dominic Derzu, Henry Mensah-Brown and Abeeku Brew-Hammond

16.1. THE INCREASING ENERGY DEMAND IN GHANA

The demand for power has increased significantly in Ghana over the years, resulting in an annual power supply crisis since 1983. Thus the country has had increasing difficulties in meeting the demand of domestic and industrial consumers, and export commitments to neighboring countries. In fact, Ghana used to be a net exporter of electricity but this situation has changed and the country is now a net importer of power from la Côte d'Ivoire. Hence there is an urgent need to look for alternative sources to widen the power generation mix in the country, while also improving the reliability of supply.

The hydropower plants at Akosombo and Kpong, and the recently added thermal plants at Takoradi, together with the import complements from la Côte d'Ivoire, cannot meet the power demand for various user categories. The short supply of electricity throughout Ghana forces many small and medium enterprises (SMEs) to run expensive standby diesel generators to meet their energy requirements. The government has decided to remove all the latent hurdles preventing private sector involvement in the energy sector and the economy has been liberalized, resulting in an influx of new investors.

Biomass resources from the agricultural and forestry sectors are readily available for energy purposes along with wind and solar resources. The limiting factors for the utilization of these energy sources include location, availability and sustainability of the resources, fuel handling and preparation, and opportunities for fuel flexibility. In addition, the technology choice and its reliability, as well as the overall economics of energy projects are particularly important, especially now that the tariff regime is being reviewed towards better economic efficiency.

This chapter looks into the potential for wood waste utilization for power generation in Kumasi, Ghana. The specific conditions for utilizing wood waste from wood-processing industries in cogeneration are presented for a project. The feasibility of the project is discussed under the framework of CDM, which also includes the boundary and baseline for the project, the carbon offsets as certified emissions

Bioenergy – Realizing the Potential

Table 16.1. Estimates of volume of residues from wood-processing activities in Ghana

Type of wastes	% of the total log input	Volume in m^3 SWE
Off-cuts	20	200 000
Bark, slabs and edgings	20	200 000
Sawdust	15	150 000
Total	55	750 000

Source: Kumasi Wood Waste Cogeneration Feasibility Report, 2001.

reductions (CERs) expected, and the outstanding issues that need to be solved to allow full implementation of the project. The project seeks to make use of abundant resources currently perceived as a nuisance to the environment, while also offering an opportunity for power generation.

16.2. AVAILABILITY OF WOOD WASTES IN GHANA

The annual cut allowance of forests to the wood-processing industry is 1 million m^3 and the Forestry Commission is responsible to ensure compliance to this law. However, the limit was not seriously observed until the enforcement of the ban on chainsaw operations in the country in 2000. More than 50 per cent of the cut allowance is attributed to sawmills in Kumasi, the most industrial timber city in Ghana. In fact, of the approximately 100 sawmills established in Ghana, 67 are in the Ashanti region, of which 65 are located in Kumasi and surroundings.

The wood waste associated with sawmilling activities in Ghana is of two categories. There are forest residues (leftovers from forest cutting) and residues from the wood-processing mills. The latter group of residues is envisaged for use in the cogeneration project presented here. Experts of the forest and wood industry in Ghana classify the volumes of residues from commercial wood-processing operations as shown in Table 16.1. Of the different types of wood waste listed, only the sawdust has no competitive use.

16.3. FEASIBILITY OF A COGENERATION PROJECT IN KUMASI

The timber industry in Ghana comprised about 134 wood-processing firms in 1996 (S-B. Atakora, 1999). It is the fourth foreign exchange earner in the country after gold, cocoa and tourism, and a heavy user of both electricity and heat. The industry has been constantly troubled with power outages due to power rationing in the entire country, especially in the dry season that lasts for almost a quarter of the year.

Meanwhile, this industry generates significant amounts of waste, which represent between 55 and 70 per cent of the total log input. This waste is in the form of sawdust, edgings, off-cuts and slabs which are suitable for power generation. It is estimated that about 3.7 million m^3 of lumber was processed in 2001. This generated between 2.0 and 2.6 million m^3 of wood waste of which only a fraction was used inefficiently in boilers to heat the kilns.

There are large quantities of sawdust currently causing nuisance in practically all the wood-processing areas of Ghana. The sawdust could be used to provide the energy needed in local industries, especially those located in periurban and rural areas where the energy crunch is felt most severely. Figure 16.1 shows a "mountain" of sawdust produced just over a single weekend operation in one of the medium-sized sawmills in the district of Kumasi. Most of this sawdust is disposed off mainly through open burning or dumping in pits at the sawmills. Besides posing great environmental hazards to the sawmills and the neighborhood, this implies the loss of significant biomass resources that could be used for energy purposes.

In 1998, the Kumasi Institute of Technology and Environment (KITE) conducted a prefeasibility study to look into opportunities of cogeneration using wood waste in Kumasi. The prefeasibility study established that:

- Cogeneration using residues of the wood-processing industry is feasible and has a great potential for meeting some of the energy requirements of the country.

Figure 16.1. Mountain of sawdust at one of the sawmills in Ghana. Source: Sawdust gasification for power generation in Ghana (Derzu and Brew-Hammond, 2001).

- The resource availability is enough to sustain cogeneration plants in the larger sawmills where continuous supply is assured.
- Kumasi is an ideal location for siting biomass cogeneration plants considering that about 60 per cent of wood-processing firms in Ghana are located here.
- Electricity tariffs in Ghana are at a level that makes cogeneration very competitive.

The above revelations led to the full feasibility study of a project, undertaken in 2001 by Econergy International Corporation (EIC) of USA in collaboration with KITE, under the auspices of the US Department of Energy. The feasibility study proposed a combined heat and power (CHP) plant with an installed capacity of $3.6\,MW_e/9.3\,MW_{th}$ using steam turbine technology to be sited at the cluster of sawmills (Kaase) in Kumasi. The plant is projected to have an annual wood waste requirement of about $80\,000\,m^3$. The CHP plant is to meet the power and heat requirements of the main sawmill, which is also the provider of the bulk of the wood waste (about 70 per cent), and a nearby brewery. This project is now possible as the Government of Ghana has liberalized the energy sector for independent power production (IPP). In addition, the project is perceived to have a good potential within the emerging CDM market.

16.4. BOUNDARY AND BASELINE OF THE CDM PROJECT

The two direct beneficiary firms of the project (the sawmill providing the bulk of the wood waste and the brewery) have both access to the national grid. Thus, the project boundaries encompass the following:

- The biomass (wood waste) cogeneration plant (assumed to have zero emissions);
- The Aboadze thermal plant at Takoradi;
- The import of thermal power from la Côte d'Ivoire;
- The standby diesel generating sets at the two beneficiary firms used during power outages;
- The residual oils used in the boilers in the brewery;
- The biomass (wood wastes) burnt in the boilers of the sawmill (assumed to have zero emissions).

This project seeks to mitigate climate change by contributing to the reduction of greenhouse gas emissions. There is a looming power crisis from 2002 and the Ministry of Energy is considering bringing in emergency power barges to be run on crude oil. The imminent power crisis is due to the low water levels in the dams at

Akosombo and Kpong. These low water levels are the result of poor rainfall which is being attributed to the negative impacts of climate change.

Just as the project boundary is dynamic, so is the baseline. In the absence of the biomass CHP plant, the sources of anthropogenic emissions of greenhouse gases are:

- The hydropower from the Akosombo and Kpong dams (zero emissions);
- The Aboadze thermal plant at Takoradi running on crude oil (2003–2006);
- The Aboadze thermal plant at Takoradi running on natural gas from the proposed West African Gas Pipeline (2006–2020);
- The import of thermal power from la Côte d'Ivoire run mainly on natural gas;
- The standby diesel generating sets at the two beneficiary firms used during power outages;
- The residual oils used in the boilers of the brewery;
- The biomass (wood wastes) burnt in the boilers of the sawmill (assumed to have zero emissions).

The issue of the timing of the emergency power additions are also applicable to the baseline scenario here. In addition, there is also the issue of the timing of the West African Gas Pipeline Project.

16.5. CERTIFIED EMISSION REDUCTIONS (CERs)

The formula for the computation of the carbon emission reductions is based on the difference between the emissions in the base case and the emissions after the implementation of the project as stated earlier in section 16.4. The baseline emissions are the carbon emissions that are likely to result in the absence of the proposed project. The basic formula for the calculation of carbon emissions is as follows:

$$\mathrm{CERs} = E_b C_b - E_p C_p$$

where: E_b = Energy produced in base case
C_b = Carbon intensity of energy in base case
E_p = Energy produced in project case
C_p = Carbon intensity of energy in project case

A summary of the calculations is shown in Figure 16.2. From the carbon analysis the anticipated carbon dioxide emissions reduction would be approximately 253 673 Mt CO_2 if the project is implemented (EIC & KITE, 2001). KITE will prepare a detailed monitoring and verification protocol, describing the specific steps to be

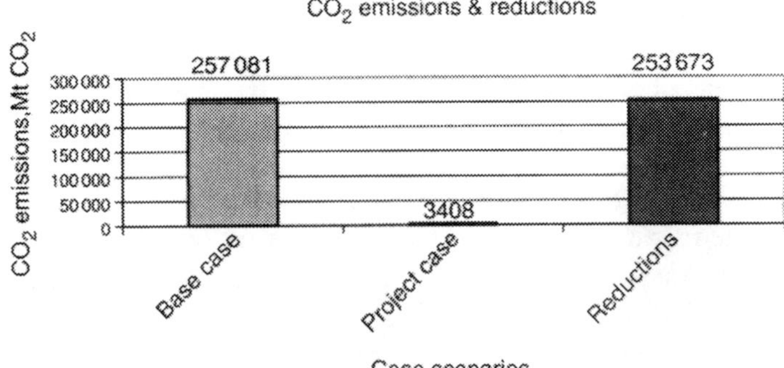

Figure 16.2. Total CO_2 emissions and total CO_2 reductions for a 10-year period, 2003–2012. Source: Kumasi Wood Waste Cogeneration Feasibility Report, 2001.

taken in the monitoring and verification process and the roles to be assumed by specific parties in this process.

It should be noted that the power imports have increased to about 7 per cent of the total country power demand and have not been captured in the carbon analysis. They could be quite significant as these imports are coming from a thermal plant using natural gas for power generation in la Côte d'Ivoire.

16.6. OUTSTANDING ISSUES

Due to the high prospects of this project as established in the 2001 feasibility report for both the economic and technical viability, the following issues are emerging:

- Recasting the 2001 Feasibility Report into the CDM Project Design Document (PDD);
- The base case scenario should be reviewed, as it is dynamic. It is anticipated that a 200 MW_e emergency diesel power barge will be made available in Tema to supplement the low hydropower production levels. The 30 MW_e decommissioned thermal plant at Tema should be eliminated from the baseline analysis;
- Critical examination of the financials with the aim of bringing down the specific power to comparative levels with other generation methods;
- Working to attract other strategic investors to complete the financing scheme;
- Upfront funding from the estimated CERs to quick-start the project implementation.

For a project such as this one, it is essential to ensure the commitment of the two main private interests to the project, that is, the timber company and the brewery.

In addition, the terms of contract for trading the excess electricity are to be defined with the Public Utility Regulatory Commission (PURC). Finalizing the rates and costs associated with electricity sales is urgent because they have repercussions for the financial analysis. KITE will revise the financial analysis based on the cost of equipment, rate and cost negotiations with PURC, rate discussions with the brewery, and revise financing assumptions.

Scenario analyses will be carried out to explore a variety of potential project conditions, as an input to project design. These will include estimations of the level of investment required to make the project feasible, the electricity sales and purchase rates required for economic feasibility, and the pros and cons of delivering steam to the brewery. An Environmental Impact Assessment is also being prepared. Forestry practices of the main contributors of sawdust to the project are being verified to guarantee the sustainability of the whole production chain.

The proposed project seeks to convert a waste product with negative environmental impact (sawdust) into a highly demanded product, which has the potential to help generate income for industrial development (electricity) and poverty reduction (employment). Today, sawdust is considered as an environmental nuisance and largely goes unutilized. During power outages, all the big sawmills that produce the bulk of sawdust run expensive standby diesel generator sets to produce electricity for their production processes. Using waste sawdust to generate electricity for these sawmills is therefore an innovative response to local needs, and a contribution to climate change mitigation.

Finally, it should be mentioned that, besides its contribution to sustainable development and remarkable reductions in carbon dioxide emissions, this project has a tremendous potential for replication. The situation found in Kumasi is by no means unique and can be observed in many other developing countries, where similar projects are also likely to be economically feasible.

REFERENCES

Atakora, S-B. & Brew-Hammond, A. (1999) Co-generation from Wood-Processing Industry Residues in Ghana, Paper presented at Biomass Conference of the Americas, August/September 2000, Oakland, California.

EIC & KITE (2001) Kumasi Wood Wastes Cogeneration Plant, a Feasibility Report, Econergy International Corporation (EIC) & Kumasi Institute of Technology and Environment (KITE).

Part V

Meeting the Challenges and Making a Difference

Chapter 17
Bioenergy – Realizing the Potential Now!

Semida Silveira

17.1. BEYOND THE BARRIERS TO BIOENERGY UTILIZATION

Bioenergy is the most important renewable energy source used in the world today. It took time for this to be fully recognized which is reflected in the fact that, only recently, efforts have been made to capture its importance in official statistics. Countries in different parts of the world have become more aware of the biomass potential they possess and the ways in which it can be used to satisfy modern energy needs and promote development. As a result, the status of bioenergy as an alternative solution to meet energy demand and mitigate climate change has improved.

We can be optimistic and say that we have reached a turning point. But we should not be too optimistic as to believe that the future of biomass is given. Though we have many scenarios that capture the biomass potential and indicate the importance it can reach, we still lack a road map to take us there. Despite all the identified advantages, bioenergy utilization has increased at a modest rate. Renewable energy technologies are not competing on a level playing field due to subsidies to conventional technologies, and disregard of negative impacts of fossil-fuel-based energy.

In this book, we have looked into ways through which the bioenergy potential can be realized. We have seen how countries at different levels of development and with different endowments can find ways to benefit from local biomass resources for both energy provision and sustainable development. Biomass resources can be enhanced through action in forestry and agriculture. Energy-related activities can not only provide complementary income but can also improve the competitiveness of established activities in these sectors. We have looked into policies, technologies and creative ways to foster collaboration among countries towards the implementation of global and local social and environmental agendas. Concrete examples of bioenergy utilization and continued activities of research and development help to understand and disseminate bioenergy solutions, improving technical expertise, managerial practices and markets.

This chapter finalizes the book with some reflections about issues that need further exploration and research in the short and medium run, and which are

Bioenergy – Realizing the Potential

necessary in the process of promoting bioenergy globally. The discussion on trade-offs in the next section brings considerations related to environmental conservation, on the one hand, and development and social equity, on the other hand. How can these considerations be more constructively linked when dealing with energy-related goals and the application of sustainability principles? In the following section, the integration of systems discussed in previous chapters is further explored in the context of markets. Finally, we end with a reflection on the link between the local and the global solutions, and the role that developing countries may play in global bioenergy solutions.

17.2. FINDING COMMON GROUND TO UNDERSTAND AND DEAL WITH TRADE-OFFS

There are a number of trade-offs to be addressed when it comes to the choice of an energy path or even a specific energy project. For example, trade-offs may relate to immediate needs and scarce financial resources versus long-term societal objectives; environmental impacts versus social and economic gains; global and national societal and environmental gains versus cost allocation. The importance attached to each of these trade-offs may vary in time and place. However, the basis for discussions about energy systems should be the need to guarantee the overall sustainability of human and natural systems.

Thus we need to go beyond the idea of trade-offs which gives a sense of conflict, and move towards the identification of what is essential in terms of energy provision. The World Energy Council has defined the three pillars of sustainable development in the context of energy as being accessibility, availability and acceptability of energy services. Accessibility is a challenge particularly focused on the need to provide 1.6 billion people in the world with modern energy services. Availability is related to the adequacy, reliability and quality of the energy supply. Finally, acceptability has to do with issues of economic affordability, as well as social and environmental impacts (WEC, 2001). This three-pillar concept contributes to highlight major foundations for future development of energy systems, going beyond the technical and supply orientation that used to characterize their expansion.

The benefits of accessing modern energy services are well known, and they have served to justify large investments in energy infrastructure. In fact, the focus on the benefits of energy provision added to little preoccupation with environmental impacts contributed to the rapid economic development observed in the past. The awareness about the full costs of achieving broad energy access is much more recent. For example, we are now aware that more than 40 per cent of lead emissions, 85 per cent of sulfur emissions and 75 per cent of carbon dioxide emissions originate from fossil fuel burning, with significant implications for the environment and health. These impacts and costs cannot be ignored in a society that aims at sustainability.

Today, we find ourselves entangled in a well-established infrastructure heavily dependent on nonrenewable resources. This infrastructure has been reliable because the technologies to deploy and use fossil fuels are mastered, synergies exist with other industries, and there are markets for these fuels operating internationally. It should not come as a surprise that fossil-fuel-based technologies have been developed and continuously improved to compete commercially, as they have been the focal point for such a long time.

There is also an established practice of subsidizing fossil fuels either directly as a way to retain jobs, for example, or by not internalizing the full costs implied in their deployment and utilization. This contributes to the strong competitiveness of fossil fuels and allows a continuous expansion of their use as if there were unlimited availability of these resources, and as if there were no limits to the resilience of our ecosystems. But fossil fuel markets are full of imperfections, particularly when it comes to oil, and affected by limitations of resource availability, and geopolitical complexities that increase price volatility with negative effects on the world economy as a whole.

In addition, previous practices contribute to the belief that energy is cheap. It gives the impression that the renewable technologies are far too expensive and not capable of competing commercially in an open market. In fact, energy has never been cheap; only we are not used to paying the full price for harnessing and using energy. The most obvious expression of the accumulated costs for energy used in the past is the increasing concentration of greenhouse gases in the atmosphere, most of which comes from fossil fuel burning. The costs of shifting energy systems towards renewable alternatives are not simply the costs of developing new technologies and creating markets for those, but also are the costs of shifting towards new infrastructure systems and paying accumulated rents.

While we are more aware of the impacts and costs of energy systems, it is important not to be trapped by them. For example, we could be trapped by the beaten track and continue the expansion of nonsustainable systems due to the political, institutional and economic difficulties implied in changing course. It could be actually claimed that this is partly happening as we watch the use of nonrenewable resources advance more rapidly than justifiable because decision makers are reluctant to set new directions and because established structures are slow in implementing change. Thus we can be trapped by the difficulties to motivate good solutions to the general public, and gamble on our future instead. We need to be clear about what the trade-offs are when it comes to energy deployment and use. We need to improve understanding about the issues involved, find ways to extract synergy benefits of choosing renewable paths, and highlight them as a way to obtain broader support for change.

When energy accessibility is considered in the context of the affordability of poor populations, economic and social trade-offs are imposed. Energy services have to be affordable. A major question in many developing countries is how to mobilize financial resources to provide energy under conditions of volatile demand due to uneven affordability, and uncertain economic returns. This requires better distribution mechanisms which have to be developed both at country level and in cooperation with the international community. Only by making developing countries part of the solution will we be able to deal with the energy and climate change debt. Bioenergy provides a road in this direction.

Using natural resources to generate social benefits is part of the sustainability concept provided resilience levels are observed, and the options left to future generations are respected. In any case, trade-offs have to be made in terms of managing natural, financial and human resources. Appropriate methodologies need to be further developed for the appraisal of energy systems which capture local and global potential gains, and which are linked to proper international cooperative regimes that help foster the most desirable solutions. The global climate change agenda is a promising new platform whereby renewable technologies can receive support to gain new markets. Bioenergy is an attractive alternative at hand in this context as discussed throughout this book.

Besides being renewable, bioenergy can bring about many environmental benefits, including the recovery of degraded land, reduction of soil erosion and protection of watersheds. If properly managed, bioenergy can be CO_2 neutral which makes it particularly attractive as a climate change mitigation option. Bioenergy may bring about significant socioeconomic benefits in the form of rural employment and positive impacts on local economies. Thus bioenergy is not only attractive from the environmental point of view, but also provides socioeconomic advantages for both developed and developing countries (Woods and Hall, 1994).

This is not to say that bioenergy is totally free of controversy. Large expansion of monoculture, competition for land and water, and quality of soils are some of the major issues related to further development of bioenergy which shall gradually become more correlated to the evaluation of its benefits. While the possibility to use local and regional potential for bioenergy is a great advantage, the transformation of biofuels into commodities and the formation of international markets shall be determinant on the extent to which bioenergy will become a major modern energy source in the next few decades. The formation of biofuel markets will benefit developing countries which, in general, have favorable conditions for growing biomass. On the other hand, the formation of these markets will have to deal with established interests in agriculture and forestry sectors of industrialized countries, requiring innovative policies.

To be able to explore the benefits of bioenergy at full scale, we need a common dialog to try and understand the local and global trade-offs as they are perceived by different interest groups. Only then will we be able to find ways for combining top down and bottom up approaches to promote the technological and social transitions of energy systems that are needed, not least to solve the climate problem. At the end, it is a matter of bridging part of the technological and social divide that we have in the world today. In this context, bioenergy provides an alternative to tackle the energy divide while also contributing to development at large. A more open dialog would allow a clear evaluation of the trade-offs implied, avoiding the simplistic and often applied dichotomy of immediate local social benefits versus long-term global environmental gains which hampers change towards sustainable development.

17.3. COMBINING POLICIES, TECHNOLOGY AND MANAGEMENT TO DEVELOP INNOVATIVE MARKETS

Improvements in resource allocation and efficiency of our energy systems are needed to attain sustainability. How can we obtain increasing utility value for a much larger number of people with shrinking resources? Energy systems produce commodities that are an essential part of industrial production systems and people's lives all over the globe. Thus energy systems need to be reliable and affordable, and transformation of their organization should not put security of supply at risk.

Energy systems are complex, composed of a variety of technologies for energy generation, distribution and use which can cause significant impacts. Even if individual technologies can sometimes be uncomplicated, the internal logistics of energy systems can be quite intricate. Single technological improvements are seldom sufficient to accomplish the extent of efficiency improvements needed. It is at systems level, in which various specific technologies are included and resources well managed, that such improvements can be achieved. Dematerialization and better allocation of financial and human resources are obviously needed.

The various dimensions of energy systems, including technical, logistical, institutional and end-use aspects, are reflected in the infrastructure built over the centuries. Intra-system relations have also evolved, providing both opportunities and constraints to the renewal of the energy infrastructure. Opportunities, because it is at the systems level that major innovations can make most difference; constraints, due to the usual sectoral approach to management and innovation. In addition, established systems generate jobs, economic output and social welfare, and any disruption of this balance is bound to create protest, unemployment and diseconomies.

The decentralization of ownership and management in infrastructure sectors has gradually led to institutional changes. Electricity and heat markets have been created. Besides increasing competition, energy companies have to respond to stiffening environmental regulations, changing service demands, and new technological choices. This forces companies to define innovative business strategies. Considerable attention has to be given to the size and quality of the demand, and there is now a stronger focus on the economic and financial dimensions of the business in contrast with the technological options at hand. This has contributed to increased efficiency within the sector but has not yet proven to offer the dynamics needed to innovate the sector at a systems level.

The political role has changed from planning the expansion and operation of the technical system per se to regulating and guaranteeing the overall balance of the system, and designing policies to foster the development of sustainable energy systems. Within this new context, new energy markets are maturing and energy infrastructure is evolving as a result of decentralized decision-making based on the given policy frameworks. If policies are ineffective, there is a risk that decisions on investments and calculations on returns become very short-sighted, leading to a less than optimal overall result. Governments need, therefore, to orchestrate energy policy in such a way that good ground is given not only for new investments but also for innovation.

The new context of energy markets is conducive to the introduction of renewable energy options and the inclusion of users as important actors in the operation and development of energy systems. Obviously, they have always been important, only the traditional engineering view of energy systems used to focus on the large-scale technological solution, often forgetting to try and understand the motivations and practices of the users. We have broadened the considerations on resource availability, technology choice and reliability and are now asking questions about acceptability, cost return and profitability. More attention to the users is necessary in a competitive market thus hopefully also resulting in better service provision.

Certainly, energy research and planning have taken a new direction due to a broadening of the energy systems perspective to include the human dimension in terms of behavior and their role in the formation of energy service markets. However, conceptual changes are not enough. The present conditions can simply lead towards further development of conventional technologies in a short-sighted market approach. In contrast, markets can be used to promote innovative energy systems if the policy framework and incentives are there to help direct the responses of markets and users.

Markets have to be created for new alternatives which are considered desirable by society but which will not easily take off unless a policy framework is put in place to promote them. IEA (2003b) talks about the three perspectives through which we analyze market formation, these being (1) the research, development and deployment perspective which focuses on innovation and industrial strategies; (2) the market barriers perspective which is the economist's perspective and focuses on decisions made by investors and users; and (3) the market transformation perspective which looks into the whole chain from production to use. IEA concludes that the three perspectives are complementary and they are all needed to help define good policies that can transform visions into practice through the discipline of markets.

Within well-functioning markets, incentives for innovation have to be created, for example to promote better resource management and increase energy efficiency. Yet there are potential synergies which may not easily occur because it would require efforts that are marginal to the core activities of the industries affected. Since the potential for efficiency improvement of each single innovation is limited, optimization at the unit level only gives relative gains compared with the gains that could be reached at a systems level (FRN, 1998). This is very relevant to consider in the context of climate change mitigation. Most greenhouse gas emissions come from the energy system and significant changes will have to take place in the way we organize energy infrastructure as a whole and in each unit. However, synergy effects exist within other sectors which affect energy demand and use. The scale of emissions reduction needed will require a broader view of possibilities for systems integration, for example, considering how urban and regional spaces operate.

There is a great stock of knowledge in the energy sector about the potential for raising efficiency. A number of measures and technologies that could contribute towards reducing greenhouse gas emissions are also known. But how will improved material efficiency affect energy intensity in industry and finally affect the demand for energy and transport, for example? How will information technology affect the demand for transport, the urban structure, and the use and demand for energy? The answers to these questions are not trivial and markets cannot provide them. The answers lie in innovative thinking as much as in technological innovation. Increased understanding of the interplay between the systems that compose our economy is needed to identify and implement the innovations that will lead to breakthrough.

17.4. GLOBAL SOLUTIONS NEED LOCAL SOLUTIONS – IMPLEMENTING STRATEGIES FOR SUSTAINABLE DEVELOPMENT AT PROJECT LEVEL

Development strategies reflect major principles and goals and give direction and guidelines for reaching societal objectives. They should also provide road maps or a development trajectory reflecting consensus around major goals. Strategies need to be constructed with the global context in mind but should build strongly on the specific conditions found in the country or region in question, be they strengths or constraints that have to be addressed in the development process. Finally, strategies are translated into specific policies and projects, which reflect both broad and specific objectives of society. Simplifying, we could see strategies linked to long-term vision while policies are designed to shape processes towards that vision. Projects translate strategies and policies into action – it is the implementation phase (Silveira, 2004).

Often, changes in environmental quality are justified in terms of social and economic gains. Difficulties arise when we try to quantify losses and gains, and identify who the losers and winners are in each case. The quantification of resilience levels and how much environmental damage is acceptable given a certain level of return is subject to different methodologies and value scales. A project could have environmental impacts beyond what seems justifiable in terms of the social gains accrued. But if the impacts are global and perceived as abstract and the gains are immediate and quantifiable in terms of jobs and economic growth, how can various objectives be conciliated, who shall pay, who shall gain and at what point in time?

Bioenergy utilization and its impacts can be seen from both local and global perspectives in very diverse socioeconomic contexts. The traditional use of biomass, for example, can strongly characterize the way a household operates in a poor rural area. Collecting biomass takes a significant amount of time and the use of biomass serves to provide services such as cooking, lighting and heating. When new technologies are inserted in this context to provide more efficient and reliable energy services through energy carriers such as electricity, liquid or gaseous fluids, or processed solid fuels, we change life styles and the way households and communities operate. We create new functions and demands for energy services that were not there before, or we satisfy a potential demand in the cases where economic activities are being hampered by lack of proper energy provision. In the latter, energy provision can open a new door that helps boost development.

When we think of bioenergy utilization in a context where modern energy services are already being provided, we are also talking about a technology transition that may affect life styles. Nevertheless, our primary initial focus is with the techno-economic transition per se and how we can put new technological solutions, and perhaps also services, into place with minimum disturbance in the quality of services provided. It is a matter of finding the right entrance for the new solution either by simply shifting smoothly towards a new solution or by adding a new service dimension as a way to motivate the change.

Bioenergy does offer many new service dimensions that are appreciated by society. While the bioenergy solutions can be engineered into existing energy systems, a broader use of bioenergy requires that it is also visualized as an integrated part of the logistics of other production systems. And this is where the real complexity of the bioenergy systems lies at the moment. The existing knowledge base is enough to put markets into operation as the experiences of various countries such as Sweden,

Austria and Brazil indicate. Now is the time for gaining momentum. This can be accomplished by identifying ways of replicating successful experiences in new formats as exemplified here in Chapters 4, 7 and 8; improving the technologies or rethinking them as exemplified in Chapters 6 and 9; improving economic efficiency by linking bioenergy solutions more closely with other systems as discussed in Chapters 3, 5 and 10.

We need the strategic thinking of public and private policy makers to reframe the platforms on which bioenergy will flourish, that is, where specific projects can be visualized, engineered and realized. A lot can be done at different project scales when it comes to bioenergy. It actually allows for a revolution in face of the opportunities available for the integration of various production systems, traditional and modern. Basically, one can only think of oil prospection and exploration as a major enterprise. In contrast, you can think of cogeneration and ethanol production in terms of the opportunities to link small- and medium-sized companies successively operating at regional, national or international levels. Within a context of well-developed energy systems, bioenergy provides a choice that can be fully integrated into existing infrastructure systems without major disturbances and, actually, with major gains.

In fact, many bioenergy technologies for power generation are advantageously used in small-scale units with further advantages in terms of location flexibility. This allows for a decentralization of the power production, which is well in line with the restructuring of electricity markets, the objectives of sustainable development and investment constraints for large-scale projects. In addition, energy infrastructure can be built step-wise allowing for a learning process even in remote areas.

Biomass resources are defined by land, water, light availability, and also labor, expertise and managerial capacity to organize a continuous and reliable production of biomass resources on a sustainable basis. The organization of biomass-based energy systems requires a number of institutional and technical arrangements, which, in turn, may need initial support and incentives from public organizations. Developing countries enjoy favorable climatic conditions, which make them particularly apt for biomass production and utilization in power and heat generation, as well as liquid biofuels such as ethanol and biodiesel. Some of these countries have large availability of land and could become net exporters of energy while great benefits are accrued in terms of job creation and income generation. Many countries lack the infrastructure or the managerial capacity to implement large bioenergy systems, but the opportunity to start at smaller scales shall be attractive for many.

Agroforestry and intercropping for optimizing resource potential, combining food and energy production with ancillary benefits such as pulp production, construction materials, fertilizers and environmental protection are fully possible today. In the northeast of Brazil, cattle raising is being successfully combined with eucalyptus plantations for pulp production, generating residues for many small industries such as ceramics and bakeries. The sustainable use of biomass crops and residues help alleviate pressure on natural forests and landscapes, while also generating new options to rural enterprises, with jobs and improved regional economy as a result.

The availability of biomass resources is important in determining a region's aptitude for bioenergy. Even with the formation of solid and liquid biofuel markets, a local resource base may remain desirable, at least until biofuel markets have matured and reliability is perceived as satisfactory. But there are other reasons to strive for a local and regional resource base. In Europe, for example, it will provide an opportunity for a new type of rural development. In peripheral areas, such as in the far northern parts of Europe, it offers a concrete contribution for the survival of local communities. This distributional dimension of bioenergy is strategically important, the challenge being to create mechanisms for balancing price competition in open international markets, on the one hand, with regional development objectives, on the other hand.

Organizing bioenergy production requires significant logistical solutions including transport systems and a variety of biomass producers, which need policy support to be operational and be able to compete with other energy forms in an initial phase. Technical aspects of logistics and generation may be straight forward in the sense that there are tested technologies and solutions which have been continuously improved, and can be readily applied. Yet local knowledge and adaptation is often needed to get projects off the ground on a sustainable basis, building a system of significant size that can bring energy services and other benefits to society, while also providing a good business base for the actors involved.

Strong focus on projects rather than systems may lead to suboptimization. Therefore, it is good if bioenergy projects can be pulled together by platforms of action. Table 17.1 illustrates the direction that some platforms of action are heading where bioenergy has a role to play. These platforms can be used to promote bioenergy projects in a context that is broader than the project itself thus enhancing its value even beyond the provision of energy services.

The need for policy coordination among different sectors of the economy delays the introduction of bioenergy even in countries where the potential is very high. Thus innovative projects, supported by incentives and capacity building are needed to boost up the knowledge and interest for bioenergy. The tasks will have to be divided and systematically implemented in order to make it possible and manageable for even very poor countries to accrue the benefits of this knowledge base and, hopefully, also of expanding biofuel markets.

Government organizations need to work closely with private actors in defining demonstration projects and specific incentives to foster the formation of markets. This includes allocating investments for projects that are financially risky but which have an important role in market demonstration. In addition, governmental agencies

Enhancing the value of bioenergy projects

The platform	The direction	Who leads the process and how?	Possible contributions of bioenergy projects
National energy sector strategies and policies	Sustainable development objectives/multi sectoral approach	Governments, providing policy framework	Improvement of natural resource management Better use of rural and urban biomass residues Regional development (e.g. through job creation, income and tax generation)
Climate-related measures	Business strategies, creation of carbon markets	Governments and enterprises, finding cost-efficient ways to mitigate climate change, also through public–private partnerships (PPP)	JI, CDM projects linking local and global gains and promoting sustainable development Introduction of modern and efficient technologies
Technology concepts and solutions	Clean production processes and chains	Enterprises, incorporating research and development of products and processes	Production and use of heat and power, and liquid fuels for global markets Improvement of management and production processes towards more efficiency
Energy efficiency measures	Innovative technological and managerial systems	Enterprises, using energy efficiency as a means to enhance the overall performance of the firm	Linking core activities with complementary biomass-related activities Logistics in biomass production, collection and transporting
Project design	Energy systems implementation	Governments, with a robust policy framework; together with enterprises applying corporate social responsibility	Regional and global market solutions based on biomass production and utilization Globalization of solutions and commodity markets
Research	Standardization, market introduction of results and solutions	Researchers and entrepreneurs	Integration of innovations Decentralized solutions Link between traditional and modern production systems
Individual project value	Integrated socioeconomic and environmental gains	NGOs, creating ancillary projects to major core activities; enterprises through corporate social responsibility	Linking costs and gains of various sectors Adding social and environmental dimensions to energy provision

can assess information and make it equally available to the various actors in the market or organize procurement to push for increased efficiency and innovation. In the long run, the support for research provided by the government will be essential to guarantee a continuous development of technologies.

17.5. MOBILIZING FORCES TOWARDS SUSTAINABLE ENERGY SYSTEMS

A sustainable energy system can only evolve as a common effort of many actors. The state has a coordinating role which is accomplished through the definition of goals of common interest, and strategies and policies to coordinate efforts in the preferred direction. The strategies and decisions of energy companies to invest are strongly influenced by the policy framework provided. Corporate decisions operate within various frameworks and define strategies which have consequences for the development of energy systems at the national and international level. These implications go beyond the considerations made at national policy design which are still strong, even in the EU where efforts are being made towards policy harmonization. With increased integration of energy markets, there is also need for increased regional and global coordination, and the absence of strong commitments have been detrimental to the systems shift that are being envisaged.

We are all energy users and we play a role as a group. Users can influence technology choices as both voters and users of energy services. But we should recognize that few users are interested in energy per se. The users want reliable energy services at reasonable prices and many also want sustainable development. But it is up to public and private decision makers, engineers and experts to engineer sustainable solutions and orchestrate change.

In the past few years, a number of initiatives have been launched by international organizations, national governments, nongovernmental organizations and the private sector. For example, in 2000, FAO and IEA Bioenergy signed a Memorandum of Understanding opening for a closer collaboration around cross-sectoral activities focused on bioenergy options. This is part of efforts being made at FAO to promote cross-fertilization of work done in forestry, rural development and energy, where bioenergy takes a prominent role. Also in 2000, a new initiative was launched in the G8 meeting held in Okinawa, Japan, aimed at promoting renewables in developing countries. In 2002 in Johannesburg, energy and development were, for the first time, treated together in an effort to set common global goals for renewable energy.

This increasing interest and support certainly provides an important base for further work in the establishment of bioenergy systems. International organizations fulfil an important role in mobilizing interest, efforts and resources. Investment banks and UN organizations can contribute in assessing and disseminating information, allocating resources for demonstration projects and liaising with national organizations to design policies and projects. International organizations can help bridge information gaps between investors and technology owners, local business and policy makers to open new channels for investments and technology transfer, thus fostering also the development process.

Most of all, we have shown here that there is significant work in progress. Among the measures envisaged to deal with remaining constraints to market penetration of renewables for electricity production, including biomass, IEA suggests measures to reduce technical problems in the form of research and demonstration, policies to level the playing field for renewables by eliminating subsidies to fossil fuel alternatives and internalization of social and environmental costs of all energy forms, as well as green electricity schemes and temporary incentives to encourage investments in renewables.

We should add that research is still strongly focused on solutions in the context of industrialized countries, while more attention is needed to understand the realities and demands of developing countries and emerging economies. This should actually also be in line with commercial interests aimed at the creation of markets for new technologies. We certainly need innovation and increased efficiency of energy systems. But we also need reliability and scale and, here, the developing countries shall be good partners, particularly in bioenergy. It is time to break the mental barrier that has transformed developing countries into a world where there is large potential demand for energy but no money. Let us seriously start considering developing countries as part of the solution – a world full of renewable resources that can help create welfare. Let us realize the bioenergy potential together.

REFERENCES

ALTENER (2001) The Impact of Renewables on Employment and Economic Growth, EU report.

EEA (2002) Energy and Environment in the European Union, Copenhagen.

EU (2001) National Energy Policy Overview available at http://energytrends.pnl.gov/eu/eu004.htm.

European Commission (2002) Let us overcome our dependence, available at http://www.europa.eu.int/comm/energy_transport/livrevert/brochure/dep_en.pdf on June 02, 2005.

Hall, D. & Rosillo-Calle, F. (1993) Why Biomass Matters: Energy and the Environment, in *Energy in Africa*, Ponte Press, Bochum, Germany.

IEA (1997) Biomass Energy: Key Issues and Priority Needs, Paris.

IEA (2003a) Renewables Information, Paris, France.

IEA (2003b) Creating Markets for Energy Technologies, Paris.

Meyer, N.I. (2003) European schemes to promote renewables in liberalized markets in *Energy Policy*, **Vol. 31**(7), Elsevier, pp 665–676.

Scheer, H. (1999) *The Solar Economy*, Earthscan, London, UK.

Schipper, L. & Meyers, S. (1992) *Energy Efficiency and Human Activity: Past Trends, Future Prospects*, Cambridge University Press, New York.

Silveira, S. (2001) Tranformation of the energy sector, in *Building Sustainable Energy Systems – Swedish Experiences*, Ed. Silveira, S., Swedish Energy Agency, Eskistuna, Sweden.

Silveira, S. (2004) Systems Approaches in Development Work, in *Systems Approaches and Their Application: Examples from Sweden*, Eds. Olsson, M-O. & Sjöstedt, G., Kluwer Academic Publishers, The Netherlands, pp 237–251.

WEC (2001) *Living in One World*, London, Great Britain.

Woods, J. & Hall, D.O. (1994) Bioenergy for Development – Technical and Environmental Dimensions, FAO Environment and Energy Paper 13.

Index

Abanico Project, Ecuador, 185, 187
  Hidrobanico S.A., developer, 185, 186
  Inter-American Development Bank, 185, 186
  Inter-American Investment Corporation (IIC), finance, 185, 186
  Internal Rate of Return (IRR), 185, 186
  Netherlands Clean Development Mechanism Facility (NCDMF), 185, 186
  Power Purchase Agreement (PPA), 184, 186
Arboriculture(al), 20, 21
Austria, BMDH
  development, 51
  economics
    Biowirt dynamic model, 53
    sensitivity analysis, 52, 53
  investment costs, 52
  microgrids, 48
  operational costs, 52
  policies, 56, 57
    energy policy, 56
  sociocultural conflicts, 54
    barriers, 54
    NIMBY syndrome, 55
    syndrome of acquired depression, 55
  systemic management, 48, 55, 57
  technology performance, 48–51
  users
    companies, 48
    farmers cooperatives, 48
    municipalities, 48
  villages, diffusion, 6, 48

Bioenergy
  benefits
    environmental, 226
    socioeconomic, 226
  development, issues, 226
  integration, of
    energy systems, 14
    forest industry, 14, 32, 37, 39, 41, 45
  options, 5, 9, 12–15, 19, 36, 37, 39, 40, 98, 107, 234
    agricultural restructuring, 16
    Brazilian ethanol programme, 9
    centralized energy planning, 13, 14
    cofiring, 6, 14, 20, 125, 126, 128, 132, 134, 135, 137–139
    Combined heat and power (CHP), 20, 21, 27, 36, 38, 40, 113, 114, 116, 118–120, 122–124, 216
    integrated gasification combined cycle plants, 20
  supply, 8, 9
  sustainable development strategies, project level, 229–234
  systems
    cofiring, 14
    ethanol additives, 14
  value-added projects, 233
  waste management, 8, 37, 45
Biofuel,
  biodiesel, 15, 22, 25–28, 231
  biogas, 8, 11, 22, 24, 64
  chain logistics, 23
  CHP plant, Ena Kraft, 38
  cycles, 23
  ethanol, 4,15, 22, 25, 27, 35, 177, 201, 231
  industries, in, 41
    gasification, 41
    pulp and paper, 34, 41, 51
  markets, 14, 16, 31, 97, 154, 164, 226, 232
  natural gas, 6, 12, 88, 91, 125

Biofuel (*continued*)
pellets, 22, 42, 99, 116, 156, 159
salix, 38
trading, 36
Biomass,
availability, 9, 27
conversion technologies, 19–22
disposal, 21
district heating, fuelwood and wood residues
Austria, Finland, Sweden, 21, 153
District Heating Systems (BMDH), 47, 48, 51–55
domestic heating, woody residues
Germany, 153
energy source/uses, 10, 15, 19, 20, 21, 26
electricity/power generation, 4, 11, 21, 27
heat, 4, 10, 11, 14
households/domestic heating, 10, 21
industry, 3, 14
transport/transportation, 4, 14, 21
EU, in
district heating, 21, 27, 47, 49, 50, 52, 53, 55
single-house heating, 27
fuels, 8, 10, 11, 15
agricultural residue, 15, 21, 87, 116
animal waste, 9
biofuel, 13, 25, 42, 153
charcoal, 15, 90, 183, 189, 192, 195
energy crops, 15, 21, 23–29, 38, 107, 154
forest residue, 9, 15, 24, 27, 44, 75, 95, 214
fossil fuel, 3, 4, 7, 8, 12–15, 21, 23–25, 27, 29, 33, 34, 44, 68, 113, 124–126, 153, 169, 174, 195, 204, 209, 210, 225, 235
short crops, 15
wood fuel/firewood, 9, 15, 20, 25, 28, 96, 98, 100, 122, 123, 141, 144
gasification, 20, 22, 41, 44, 78, 127–129, 139, 144
pathways, 27, 29
potential, 4–6, 8–10, 12, 15, 16, 21, 29, 40, 41, 92, 96, 141, 144, 153–155, 164, 177, 223
technology
combustion, 22, 25, 49, 75, 88, 91, 157, 159, 160, 162, 184, 211
esterification, 22
fermentation, 22, 78
gasification, 20, 22, 41, 78, 127–129, 145, 155
pyrolysis, 22
utilization, 4, 8, 23, 24, 31–35, 92, 95, 125
global, 9
OECD, 4, 8, 10, 11
Biomass, Baltic
Baltic Sea Region Energy Co-operation (BASREC), 107
Bio2002Energy project, 107
forest residues, 95
Biomass-based power plants, Sri Lanka
benefits, 141
economics
fuel and maintenance cost, 146, 147
Internal Rate of Return on Equity (ROE), 149
plant installation cost, 146–148
energy plantations, 141–145, 148, 149
eucalyptus, 144, 145
legumes, 144
maximum usable land, 143, 144
woody biomass, 144
land alienation
and population, correlation, 143
homestead increase, 143
potential land area, 141–143
cultivation, types, 142
dry farming, chena, 142, 149
land, types, 142
technology
down draft fixed bed gasifier, 145
pressurized fluidized bed gasification, 145
wood gasification, 145

Biomass, Sri Lanka
bagasse, 141
Energy Forum, 141
primary energy source, 141
rice husk, 141
wood fuel, 141
Brazil, biomass
Brazilian alcohol program (PROALCOOL), 75, 92
sources, 75
forest residues, 75
sugarcane, 75
Brazil, cofiring
bagasse and natural gas, 126
biomass utilization, 126, 128
Cases A, B
Biomass Gasification Integrated to Combined Cycles (BIG-CC), air-blown gasifiers, 128, 129, 132
combined cycle power plants, 128
gas turbine cycles integrated to biomass gasification (BIG-GT) technology, 129
gas turbine derating, 128–131, 133, 134
gasification, syngas, 129, 131
natural gas combined cycles (NG-CC), 131, 132
Case C
Advanced System for Process Engineering (ASPEN), 129
alternative comparison, 137
feasibility analysis, 136
gas-turbine performance code, 129
Rankine cycle, 128
computational simulations, 128
gasification technology, 127
biomass-related problems, 128
heat recovery system generators (HRSG), 128–131, 135
natural gas cofiring benefits, 125, 128
feasibility study
carbon credits, 134–136
internal discount rate (IDR), investment, 132–136
Natural Renewable Energy Laboratory (NREL), 125
simulation and feasibility results, 131–136

Carbon
credits, 23
cycle, 23
sequestration, 23, 24, 42, 43, 211
sinks, 21, 23, 24, 26, 27, 29, 42, 61, 176
substitution, 23, 27
Carbon finance
Carbon Finance Business (CFB), World Bank, 179, 180
carbon funds, 179
climate issues, 180
OECD, 179
roles, 179, 180
examples, 183–186
project-based transactions, 179
project finance risks, 181
CEN, 181
country risk/sovereign and political risk, 180, 181
country risk insurance, 181
credit risk, 181
project sponsors, uncertainties, 180
renewable energy, 180
rural electrification, 180
Carbon rights, risk mitigants, 182
creditworthy off-takers, 182
Emission Reductions Purchase Agreement, ERPA, 182–187
foreign exchange risk, 182
long-term contracts, 182
risk of price fluctuation, 182
local government and actions
letter of approval (LoA), 182
Clean Development Mechanism (CDM), 7, 32, 43, 133, 170, 171, 174–177, 179, 183, 186, 187,189, 193, 196, 201, 202, 204–206, 209, 211, 213, 216, 218, 233

Clean Development Mechanism (*continued*)
bioenergy options
carbon sink, 176
strategies, 177
climate change mitigation challenges, 169
international economy and development, 169
environmental sustainability, 169
Climate Convention (UNFCCC), 169, 204
concept, 170, 171
participation, 174
NGOs, 175, 176
SMEs, 175, 176
potential barriers, 174,175
projects, 172–177
CER accreditation, 171
cycle, 172
host country approval, Designated National Authority (DNA), 172
project design document (PDD) preparation, 172, 173
requirements, 171
climate change mitigation mechanisms, Kyoto protocol, 170
aims, 170
Clean Development Mechanism (CDM), developing countries, 170
climate change, 21, 23, 24, 29
Emissions Trading (ET), 31, 32, 43, 170
fossil fuel emissions, 21
green electricity, 28
Joint Implementation (JI), collaborative target achievement, 32, 43, 107, 170
CO_2 emissions reduction, 23, 42, 43, 92, 133, 141, 153, 165, 217–219
Cogeneration project, Brazil
Brazilian Ten-year Expansion Plan, 201
mineral coal, 201
natural gas, 201
petroleum derivatives, 201
Santa Elisa, bagasse, 201–204, 206–209, 211, 212
additonality, project, 204, 205
baseline emissions benchmark, 206
carbon accounting evaluation methods, 209–211
carbon credit, 204, 211
CERs criteria, Climate Convention, 204, 205
Companhia Energética Santa Elisa, 201, 203, 204
Companhia Paulista de Força e Luz (CPFL), energy buyer, 203, 207, 208
emissions reduction, quantification, 209–211
emissions reduction mechanisms, 210, 211
high efficiency pressure boiler installation – Phase 1, 202, 203
indirect emission impacts, 211
lifetime, project, 211, 212
National Bank for Economic and Social Development (BNDES), funding, 204
quantification, baseline carbon intensity, 207
Special Agency for Industrial Finance (FINAME), funding, 204
turbo generator acquisition – Phase 2, 203, 204

Department of Environment, Food and Rural Affairs (DEFRA), UK, 27

Emissions
CH_4 26, 130, 184, 185, 189, 192–194
fossil fuels, 21
mitigating equipments, 49
continuous power control, 49
electronic combustion control, 49
flue gas condensation, 49
nitrogen gas, 24, 26
Energy
crops, 21, 23–29
eucalyptus, 10, 20
maize, 10

miscanthus, 20, 27
oleagenous, 22
short rotation coppice (SRC), 25, 27
sugarcane, 10
forests, 20, 21, 25
integration strategies, 227
energy policy regulations, 228
energy research and planning, 228
global energy market integration, 234
IEA perspectives, 228, 229
ownership decentralization, 228
policy making strategies, 231
resource optimization, 231
technology dematerialization, 227
primary mix, 19, 20, 24, 29
residues, 21, 24
three-pillar concept, World Energy Council (WEC), 224
sustainable development, 224
trade-offs, 224
vectors, 29
Environment
acid rain, 13
Agenda 21, 3
carbon, 23, 24, 42, 43
climate bonus, 43
climate change, 20, 21, 23, 26, 27, 29
levy, 27
Climate Convention, 3, 36, 78, 169
decarbonization, 12
global environmental agenda, 3
EU policies
Common Agricultural Policy (CAP), 25, 27
Italy
Biomass Implementation Programme, 28
Netherlands
fiscal instruments, 28
green funds, 28
UK
Green Fuels Challenge, 26
Renewables Obligation, 26
EU Standardization, solid biofuel
agriculture and forestry products, 157
chain, 161
classification, low emissions property, 156, 157, 159
cork waste, 158
European Committee for Standardisation (CEN) TC, 335 157, 158
Finnish Standards Association (SFS), classification, 158, 159
German Institute for Standardization (DIN), terminology, 158, 159
Netherlands Standardization Institute (NEN), chemical testing, 158, 160
Swedish Standards Institute (SIS), physical testing, 158, 160
food-processing industry waste, 157
fuel quality specifications, 155
national standards, 155
physical–mechanical and mechanical testing, 160
quality assurance, 159
straw quality improvement, 161–163
sampling and sample reduction, 160
source and type, 155
standardization of properties, 154–156
terminology standard, 158
wood pellet standard, Germany, 156
wood waste, 158

Forestry area, India, 68
Forest fuel costs, factors, 103
Fossil fuels, 3, 4, 7,12–15
coal, 3, 12
gas, 3, 8
natural gas, 6, 12
oil, 3, 12, 13
Fuelwood
Hindu Kush Himalayan Region (HKH), 67
issues, 68–72
policy, 62, 68, 70
technology, 65, 69

Ghana, energy
biomass utilization
limiting factors, 213
CDM projects, 216, 217
baseline, 216
boundaries, 216
contribution, 219
issues, 218
Certified Emission Reductions (CERs), 217, 218
formula, carbon emissions, 217
Public Utility Regulatory Commission (PURC), 219
contract trading terms, 219
wood-processing residues, 214
wood waste availability, 214
wood waste utilization, 213
cogeneration project, Kumasi, 214
Ecoenergy International Corporation (EIC), 216
Kumasi Institute of Technology and Environment (KITE), prefeasibility study, 215
sawmills, sawdust, 214, 215, 217–219
Greenhouse gas emissions (GHG), 13, 21–24, 27, 40, 43, 124, 169–177, 179, 180, 183, 185, 189, 193, 201, 205, 209, 216, 229

Hindu Kush Himalayan region HKH, energy
agricultural diversification/intensification, 63
agro-based facilities, 63
and forests, 66
biomass-dominant fuel, 63
biomass briquettes, 65
gasifiers, 65
cottage industries, for, 62–65
forest management interventions, costs and benefits, 72
forest resource
availability, 66
consumption, 66, 67
household sector, 62
Nepal, community forestry program
forest management types, 71
forestry user's group, 71
fuelwood sustainability, 71
marginalization, 71
property rights, 70, 71
rural industrialization, 63
use pattern, 63, 64
wood energy development, 69
improved cooking stove (ICS), 70
policy interventions, 70
renewable energy technology commercialization, 70

IEA (International Energy Agency), 7–9, 11, 45, 228, 229, 234, 235
IPCC (Intergovernmental Panel on Climate Change), 10, 21, 210
Iron industry, Brazil, 190
Brazilian Association of Renewable Forests (ABRACAVE), 192
Brazilian Institute for the Environment and Renewable Natural Resources (IBAMA), statistics, 192
categories, 190
differences, coke- and charcoal-based, 193
employment source, 197
socio-economic issues, 197

Kyoto Protocol, 7, 21, 23, 24, 29, 36, 43, 189

Lithuania
and Sweden, bilateral cooperation, 95, 96
forest authorities, 97
General Forest Enterprise, 98
State Forest Enterprise, 96, 98
forest management, 96
commercial thinning, 97, 101, 102, 103–106
Forest Control Units, implementation, 98

Lithuania (*continued*)
Ministry of Environment, strategies, 97, 98
new fuel collection practices, 96
precommercial thinning, 97, 102, 105, 106
private forestland, 97
Rokiskis forest study, 100, 101, 103
short rotation age, 97
technical corridors, 102, 104, 105
forest ownership structure, 97
forest restitution, 96
fuelwood utilization
district heating, 97, 98
household heating, 97
industries, 98
integrated technologies, 104–106
economic efficiency, 106
forest fuel increase, aspen, 105
production cost decrease, 103–106
Lithuanian Energy Institute, 107
Lithuanian Forest Research Institute, 101
renewables, financing
guarantees for energy saving investments, 107
tax reduction, 107
wood waste potential, 99
briquettes, 99
exports, 99
pellets, 99

Mountains
carbon sinks, 61
environmental stocks and flows, 61
forest, in
importance, 61
forest resources management framework, 72
fuelwood sector
environmental consequences, 69
policies and programs, 70, 71
strategies, 71
sustainability issues, 68, 69
sustainable development, approach, 61

Nitrogen
efficiency, 24
fertilization, 24
fixation, legumes, 144
losses, 24
emissions, 24, 26
leaching, 24
NovaGerar Project, Brazil, 184, 187
carbon credits buyer, credit worthiness, 187
carbon dioxide equivalents, high, 185
Internal Rate of Return (IRR), high, 185
methane, electricity, 184
NCDMF, 185
sanitary landfill, 184

Plantar project, Brazil
baselines, components, 193
biodiversity concerns, environmental, 196
biodiversity benefits/indicators, 197
fire suppression, 196
Forest Stewardship Council standard, 196, 198
quality monitoring, 197
business model, 198
coal substitution, 183
country risk insurance, 183
credit-risk free, 184
Development Bank of State of Minas Gerais (BDMG), 192
eucalyptus plantations, 183, 193
coke displacer, eucalyptus, 189
export pre-payment/monetization of ERPA receivables, 183, 187
GHG emissions reduction, 189
cerrado, 189, 193
methane, 189, 193
forest legislation, charcoal use, 191, 192
international property rights, emissions reductions, 195
location
Curvelo, Felixlândia, Sete Lagoas, 189
State of Minas Gerais, 189, 190

Plantar project, Brazil (*continued*)
  Prototype Carbon Fund, World Bank, 183, 189
  Rabobank, 183, 184
  territorial boundaries, 194

Renewables
  forms
    bioenergy, 14
    biogas, 8, 11
    hydropower, 3, 11
    municipal waste, 11
    solar, 7, 8, 10
    tide, 8
    wind, 7, 8, 10
  systems, 14
  technology
    fossil-fuel-related, 12
    nuclear power, 7, 12
  world energy supply, 7, 8

Solid biofuels
  advantages, 153
  barriers, 154
  chain, principle of
    basic condition, 161
    processes, 161
    quality factors, 161
  properties, 154, 156
  potential, EU, 153, 154
Sugarcane, Brazil
  bagasse
    briquetting, 87
    characteristics, 86, 87
    energy value, 90, 91
    industrial residue, 87
  bagasse, uses, 76, 77, 87, 90, 91
    agronomic benefits, 88
    ash content, 88
    boilers, 87, 89, 91
    fire wood substitute, 91
    particle size, 86–88
    vegetable oil industry, 91
  cane-burning technique
    drawback, 75
    legislation, 77, 79, 92
    reasons for prevention, 81
  cost reduction
    Geographical Information System (GIS), 79
    logistic optimization, 80
    operation research techniques, 79
    precision agriculture, 79
  cultivation, ratoon system, 78
  energy, types, 76
    generation technology, 78
  ethanol, from, 75–78, 81, 90
  green cane harvesting, 75, 76, 81, 82, 84, 92
    emissions reduction, 81
    nitrate leaching reduction, 81
    sucrose loss elimination, 82
  mechanical harvesting
    Agronomic Institute of Campinas (IAC), 78
    bagasse pelletization, 87
    National Plan for Sugarcane (Planalsucar), 78, 79
    São Paulo, 78–81, 83, 87, 91, 92
    Sugar and Alcohol Institute (IAA), 75, 78, 79
    Sugar and Ethanol Producers of the State of São Paulo (Copersucar), 78–80, 87, 88
  research and technology, 78–80
  trash
    agricultural residue, 87
    characteristics, 86, 88, 90
    class and case machines, 87
    components, 87
    energy value, 90, 91
    industries, in, 90
    recovery, 77, 78, 81, 87, 88
    UNICAMP unbaling-feeding system, 87, 89, 90
Sugarcane harvesting
  Chopper harvester, 81, 87

green cane
benefits, 81
considerations, 82
constraints, 83
management, 92
mechanization
advantages, 84
Australian chopped cane, 80
chopped cane system, Texas, 80
drawbacks, 83–85
green cane harvesting, Brazil, 81
KTP cane harvesters, 80
Push-rake, Hawaii, 80
Soldair, 80
technical barriers, 83, 86
technology
drawbacks, 87, 88
raking and windrowing of tops, 88
Sweden, bioenergy
development
biomass enhancement, 39
CO_2 emissions reduction, 42
systems integration, 41
drives, 31, 32, 35, 36
equipments
forest fuel harvesting, 44
heat pumps, 34
pellet burners, 44
integration
electricity, 35
Enköping, 38
European energy matrix, 37
internationalization, 32, 36
bioenergy options, 36
biofuel imports, 36
R&D, 36
know-how
biomass gasification, 44
forest fuel logistics, 44
mainstream alternative, 32, 37, 39
CHP, 40
district heating, 34
motor fuel, transport, 35
single-house heating, 34
policies, 31
emissions trading, 31
green certificates, 31
producer
forest industry, 33, 34
renewable resource, 33
utilization, biomass, 31–35
Swedish Energy Agency, 33, 96, 113
Sweden, energy system, 113
CHP potential, 114, 120
district heating grids, 114
GHG reduction, 124
Sweden, small-scale district heating
cogeneration of heat and power, 122
counties, Kalmar, Örebro, Västernorrland, 116, 117, 121
heat demand estimate, 114, 115
CHP, 116
geographical clusters, 114
GIS, 115
practical potential, 114
small geographic clusters, 117, 118
technical and economic feasibility, 114, 116
theoretical potential, 114–117
geographic distribution, 117
overall heat demand, 116

Wood processing industry, Lithuania
logging activity expansion, 99, 102
restructuring, 99

Printed in the United States
49478LVS00001B/11